Lecture Notes in Business Information Pr

Chun Ouyang
Jae-Yoon Jung (Eds.)

Asia Pacific Business Process Management

Second Asia Pacific Conference, AP-BPM 2014
Brisbane, QLD, Australia, July 3-4, 2014
Proceedings

Volume Editors

Chun Ouyang
Queensland University of Technology
Brisbane, QLD, Australia
E-mail: c.ouyang@qut.edu.au

Jae-Yoon Jung
Kyung Hee University
Yongin, Republic of Korea
E-mail: jyjung@khu.ac.kr

ISSN 1865-1348 e-ISSN 1865-1356
ISBN 978-3-319-08221-9 e-ISBN 978-3-319-08222-6
DOI 10.1007/978-3-319-08222-6
Springer Cham Heidelberg New York Dordrecht London

Library of Congress Control Number: 2014940936

Typesetting: Camera-ready by author, data conversion by Scientific Publishing Services, Chennai, India

Printed on acid-free paper

Springer is part of Springer Science+Business Media (www.springer.com)

Preface

This volume collects the proceedings of the second Asia-Pacific Conference on Business Process Management (AP-BPM 2014) held in Brisbane, Australia, during July 3-4, 2014. The conference aims to provide a high-quality forum for researchers and practitioners to exchange research findings and ideas on BPM technologies and practices that are highly relevant (but are not limited) to the Asia-Pacific region. Another key objective of the conference is to set up a bridge between actual industrial requirements and leading-edge research outcomes on the growth of economic rising powers of the Asia Pacific region.

As the second edition in this conference series, AP-BPM 2014 attracted an increasing number of submissions: 33 (qualified) submissions, comparable to the inaugural AP-BPM conference held in Beijing, China, last August. These submissions reported on up-to-date research findings of scholars from 12 countries (China, Korea, Australia, Indonesia, Malaysia, Japan, India, The Netherlands, Italy, Portugal, Germany, and USA). After each submission was reviewed by at least three Program Committee members, nine full papers were accepted for publication in this volume of conference proceedings (i.e., 27.3% acceptance rate). These nine papers cover various topics that can be categorized under four main research focuses in BPM, including process mining (three papers), process modeling and repositories (two papers), process model comparison (two papers), and process analysis (two papers).

In addition, another 11 submissions were accepted as short papers. They are not published in this volume but seven of them were finally included in the conference's scientific program, as the topics and research findings of these papers were considered interesting for discussion at the conference.

This year the conference also featured with two invited keynote presentations. On the first day, Hyerim Bae, Professor at in the Industrial Engineering Department at Pusan National University, Korea, talked about a number of activities that can be carried out, using the process models discovered from the event logs, to deliver valuable inputs for industry. An extended abstract of this keynote is included in the proceedings. On the second day, Michael Rosemann, Professor and Head of the Information Systems School at Queensland University of Technology, Australia, contributed with inspiring insights into proposals of three future research and development directions for BPM academics and professionals. A full paper of this keynote is included at the beginning of the proceedings.

We would like to thank the Program Committee members and the external reviewers for their thorough reviews and discussions of the submitted papers. We express our gratitude to other conference committees as well, especially to the general chair, Arthur ter Hofstede, and the Steering Committee for their valuable guidance, to the organization chair, Moe Wynn, and other staff at Queensland

University of Technology for their attentive preparations for this conference, and to the publicity chairs, Artem Polyvyanyy, Minseok Song, and Zhiqiang Yan, for their efforts in publishing conference updates and promoting the conference in the region. Last but not least, we are thankful to the authors of the submissions, the keynote speakers, the presenters, and all the other conference participants, because the conference could not be held without their contributions and interest.

July 2014

Chun Ouyang
Jae-Yoon Jung

Organization

AP-BPM 2014 is organized in Brisbane, Australia, by the Queensland University of Technology.

Steering Committee

Hyerim Bae	Pusan National University, Korea
Arthur ter Hofstede	Queensland University of Technology, Australia
Jianmin Wang	Tsinghua University, China

International Advisory Committee

Arthur ter Hofstede	Queensland University of Technology, Australia
Hyerim Bae	Pusan National University, Korea
Jianmin Wang	Tsinghua University, China
Pingyu Hsu	National Central University, Taiwan
Riyanarto Sarno	Sepuluh Nopember Institute of Technology, Indonesia

General Chair

Arthur ter Hofstede	Queensland University of Technology, Australia

Program Chairs

Chun Ouyang	Queensland University of Technology, Australia
Jae-Yoon Jung	Kyung Hee University, Korea

Organization Chair

Moe Thandar Wynn	Queensland University of Technology, Australia

Publicity Chairs

Artem Polyvyanyy	Queensland University of Technology, Australia
Minseok Song	Ulsan National Institute of Science and Technology, Korea
Zhiqiang Yan	Capital University of Economics and Business, China

Program Committee

Saiful Akbar	Bandung Institute of Technology, Indonesia
Majed Al-Mashari	King Saud University, Saudi Arabia
Joonsoo Bae	Chonbuk National University, Korea
Hyerim Bae	Pusan National University, Korea
Jian Cao	Shanghai Jiao Tong University, China
Namwook Cho	Seoul National University of Technology, Korea
Lizhen Cui	Shandong University, China
Zaiwen Feng	Wuhan University, China
Arthur ter Hofstede	Queensland University of Technology, Australia
Marta Indulska	University of Queensland, Australia
Jae-Yoon Jung	Kyung Hee University, Korea
Dongsoo Kim	Soongsil University, Korea
Kwanghoon Kim	Kyonggi University, Korea
Minsoo Kim	Pukyung National University, Korea
Raymond Lau	City University of Hong Kong, SAR China
Chun Ouyang	Queensland University of Technology, Australia
Helen Paik	The University of New South Wales, Australia
Artem Polyvyanyy	Queensland University of Technology, Australia
Punnamee Sachakamol	Kasetsart University, Thailand
Shazia Sadiq	University of Queensland, Australia
Lawrence Si	University of Macau, SAR China
Minseok Song	Ulsan National Institute of Science and Technology, Korea
Jianmin Wang	Tsinghua University, China
Lijie Wen	Tsinghua University, China
Ingo Weber	NICTA, Australia
Raymond Wong	The University of New South Wales, Australia
Moe Thandar Wynn	Queensland University of Technology, Australia
Bernardo N. Yahya	Ulsan National Institute of Science and Technology, Korea
Zhiqiang Yan	Captial University of Economics and Business, China
Jianwei Yin	Zhejiang University, China
Sira Yongchareon	Unitec Institute of Technology, New Zealand
Yang Yu	Sun Yat-Sen University, China
Liang Zhang	Fudan University, China
Yang Zhang	Beijing University of Posts and Telecommunications, China

Additional Reviewers

Simon Dacey
Guosheng Kang
Veronica Liesaputra

Keynotes

What We Can Do with Process Models after We Discover Them from Event Logs

Hyerim Bae

Industrial Engineering Dept., Pusan National University, Busan, Korea
hrbae@pusan.ac.kr

Extended Abstract

Process mining is a process management technique that allows us to analyze business processes based on event logs [1]. Since process mining was first introduced, its main use has been the discovery, from event logs, of correct high-fitness process models [2]. As the results of early process discovery trials, many process mining algorithms, such as α-Algorithm, Fuzzy mining, Heuristic mining, and genetic algorithm, have been developed: these help us to find good process models that reflect event logs precisely.

However, in real business environments, people show more interest in the practical use of process models discovered from event logs. For example: they want to find higher levels of knowledge, diagnose their system, or find the causes of problems in their company and solve them. In this keynote address, which is based on the experience of actual process mining projects in Korea, I will talk about what we can do with process models after discovering them from event logs.

To illustrate our methods, we use two sets of event logs: one generated while containers are handled in a container port [4, 5], and the other generated while ships are assembled in a ship yard [3]. After we discover process models from these two sets of huge logs, there are three things that we can do with them.

1) Model-based real-time monitoring of process instances

Many companies think of their processes as important assets. And yet, they are unsure if process instances exactly follow the defined model. Even if we assume that a process model discovered from event logs reflects a real process well enough, there are always requirements for real-time monitoring of process instances.

Fig. 1 is a screen capture of a monitoring system that shows the current block-movement status in a ship yard. For managers, the system indicates each block's status by showing, in the discovered process model, the location of its current activity. The system also indicates the path, or work flow, of each block.

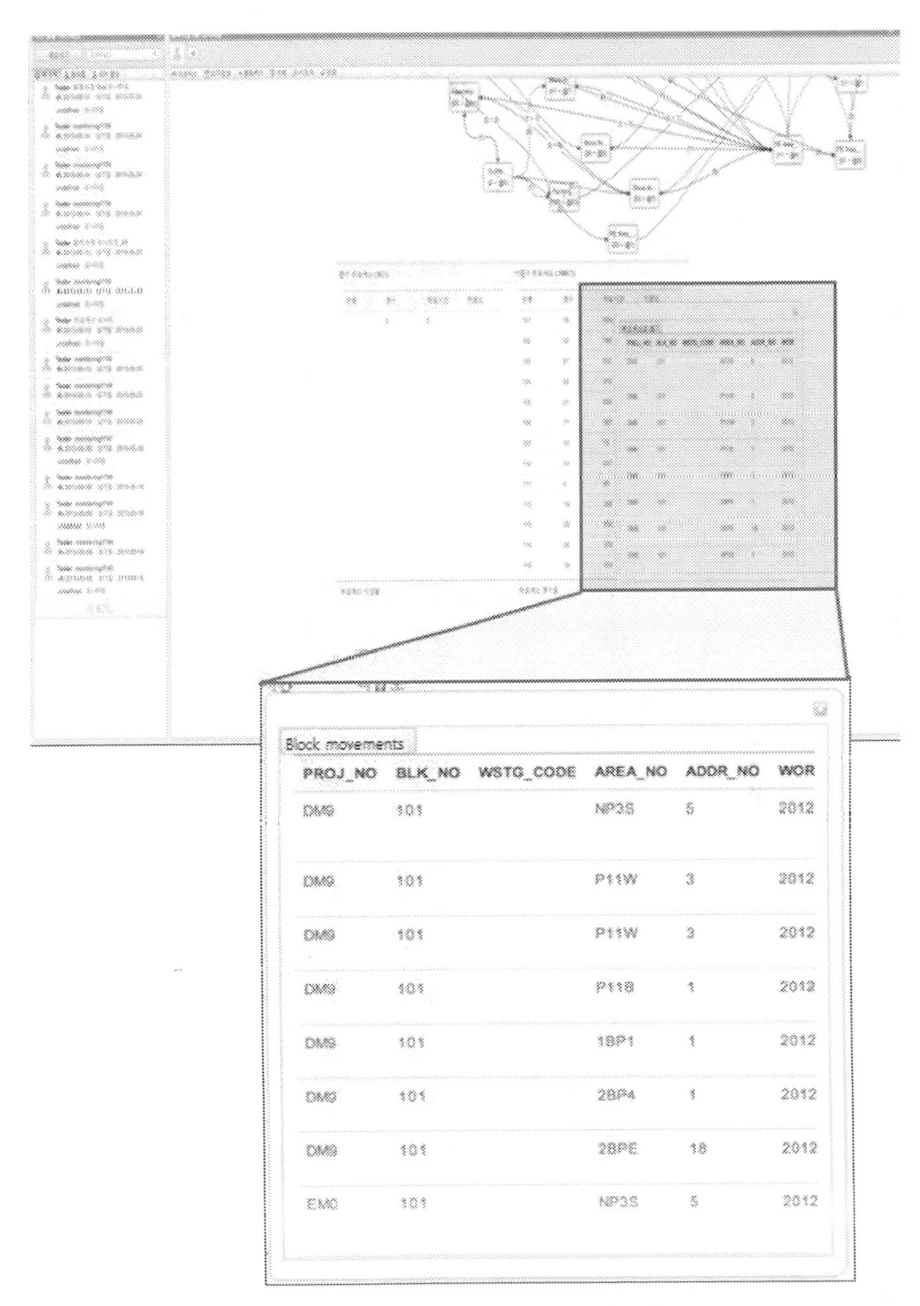

Fig. 1. Block-movement monitoring

2) Problem solving based on a process model

We can find and solve problems using a discovered process model. A central function of problem finding and solving is to detect work flows that do not exist in the process model. In fact, sometimes we can find a work flow that managers were not aware of. In that case, we can notify managers of it. For the purposes of problem solving, we can also carry out time-gap analysis. The time gaps between two arbitrarily chosen nodes can be listed in descending order. This way, we can see that the instance with the longest time gap has a problem. Also, as an advanced functionality of problem solving, multi-dimensional analysis can be introduced.

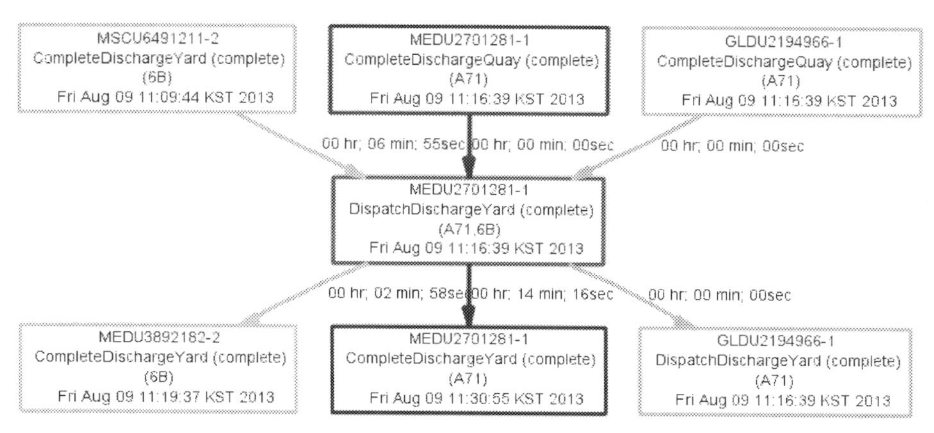

Fig. 2. Multi-dimensional process analysis for problem solving

Fig. 2 shows a multi-dimensional process view of container-handling events. After a user isolates a suspected problematic event with the red box, the system shows the previous and next events. The gray boxes show the event flow of the 6B block in the container yard— they indicate both where the red event has been dealt with, and where it will be dealt with. The blue boxes show the previous and next work events of the container. The sky-blue boxes show the work flow of the crane that handled the container. On each arc between two events, the time gap between two completed activities is represented. The number of previous and next events can be increased by user request. Using this multi-dimensional process view, we can infer what attribute causes the delay of an event.

3) Prediction of process result based on process model

A process model can be used to predict the result of process execution. In this keynote address, a prediction method that uses a Bayesian Network will be presented. A process model discovered from an event log can be understood as representing the dependency relations among activity nodes. That is, if two activities are connected by an arc, the execution of the previous activity influences the result of the next activity. In this regard, we can generate a Bayesian Network from an event log, and, using the network, we can predict the result of process execution with conditional probability.

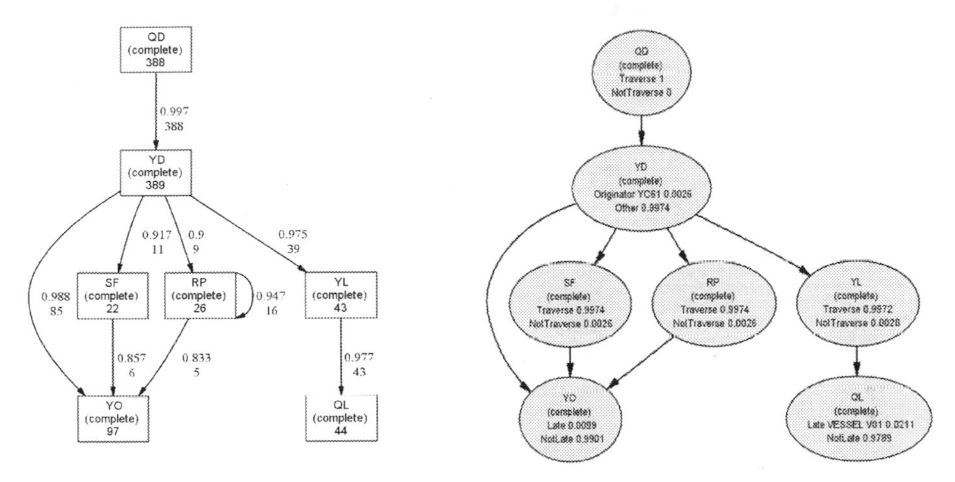

Fig. 3. Process model and Bayesian Network generated from event log

References

1. van der Aalst, W.M.P.: Process Mining: Discovery, Conformance and Enhancement of Business Processes. Springer, Berlin (2011)
2. van der Aalst, W.M.P., Weijters, A., Maruster, L.: Workflow Mining: Discovering Process Models from Event Logs. IEEE Transactions on Knowledge and Data Engineering 16(9), 1128–1142
3. Park, J., Lee, D., Bae, H.: Event-log-data-based method for efficiency evaluation of block assembly processes in shipbuilding industry. ICIC Express Letters: Part B: Applications 5(1), 157–162 (2014)
4. Sutrisnowati, R.A., Bae, H., Park, J., Ha, B.-H.: Learning Bayesian network from event logs using mutual information test. In: The 2nd International Workshop on Knowledge-Intensive Business Process, KiBP 2013 (2013)
5. Sutrisnowati, R.A., Bae, H., Park, J., Pulshashi, I.R.: Knowledge discovery of port logistics process using partial directed acyclic graph composition. In: The 17th International Conference on Industrial Engineering: Theory, Application and Practice, IJIE 2013 (2013)

Proposals for Future BPM Research Directions

Michael Rosemann

Queensland University of Technology
Information Systems School
2 George Street, Brisbane Qld 4000, Australia
m.rosemann@qut.edu.au

Abstract. Business Process Management has substantially matured over the last two decades. The techniques, methods and systems available to scope, model, analyze, implement, execute, monitor and even mine a process have been scientifically researched and can be in most cases deployed in practice. In fact, many of these BPM capabilities are nowadays a commodity. However, an opportunity-rich environment and rapidly emerging digital disruptions require new BPM capabilities. In light of this context, this paper proposes three future research and development directions for BPM academics and professionals. First, Ambidextrous BPM demands the shift of focus from exploitative to explorative BPM. Second, Value-driven BPM postulates a stronger focus on the desired outcomes as opposed to the available BPM methods. Third, Customer Process Management suggests complementing the dominating internal view of BPM with a stronger, design-inspired view on the process experiences of external stakeholders.

Keywords: Ambidextrous BPM, value-driven BPM, customer process management, design-led process innovation, configurable BPM.

Table of Contents

Keynote Paper

Process Mining

Process Modeling and Repositories

Process Model Comparison

Process Analysis

Proposals for Future BPM Research Directions

Michael Rosemann

Queensland University of Technology
Information Systems School
2 George Street, Brisbane Qld 4000, Australia
m.rosemann@qut.edu.au

Abstract. Business Process Management has substantially matured over the last two decades. The techniques, methods and systems available to scope, model, analyze, implement, execute, monitor and even mine a process have been scientifically researched and can be in most cases deployed in practice. In fact, many of these BPM capabilities are nowadays a commodity. However, an opportunity-rich environment and rapidly emerging digital disruptions require new BPM capabilities. In light of this context, this paper proposes three future research and development directions for BPM academics and professionals. First, Ambidextrous BPM demands the shift of focus from exploitative to explorative BPM. Second, Value-driven BPM postulates a stronger focus on the desired outcomes as opposed to the available BPM methods. Third, Customer Process Management suggests complementing the dominating internal view of BPM with a stronger, design-inspired view on the process experiences of external stakeholders.

Keywords: Ambidextrous BPM, value-driven BPM, customer process management, design-led process innovation, configurable BPM.

1 Introduction

Business Process Management (BPM) has substantially matured over the last two decades. More than 20 years ago, Michael Hammer (1990) [1] and Thomas Davenport (1992) [2] initiated an entire stream of activity in practice and academia dedicated to the way organizations conduct their business processes. Their contributions took previous process-related work in manufacturing going back as far as Adam Smith, Frederick Taylor [3] and Henry Ford into the boardroom and made 'process' a topic and strategic design variable across industries and disciplines.

In the early 90s, BPM was largely seen as being enabled by large Enterprise Systems as they provided, pre-defined configurable processes as part of their comprehensive packages. Scheer's contributions towards a reasonable easy way to communicate a tool-supported, integrated process modeling technique (EPCs) further accelerated the uptake of BPM [4]. In addition to process-aware Enterprise Systems, dedicated workflow management systems (e.g., Staffware, Flowmark, COSA) entered the market in order to support processes largely outside such Enterprise Systems.

C. Ouyang and J.-Y. Jung (Eds.): AP-BPM 2014, LNBIP 181, pp. 1–15, 2014.

However, the radical re-design and fundamental process innovation as postulated by Hammer and Davenport faced two challenges.

First, beyond their compelling narratives, there was a very limited set of methods available in support of process re-engineering and innovation. Instead, the focus of corporate development and academic research efforts largely went into process modeling notations (e.g., BPMN), related validation and verification efforts (e.g., livelocks, deadlocks), process analysis and assessment techniques (e.g., Six Sigma, bottleneck analysis, activity-based costing) and process execution capabilities (e.g., workflow management solutions). As a consequence, the capability to specify, incrementally improve and automate processes has grown substantially. Nowadays, increasingly complex challenges (e.g., adaptive case management, exception handling, process similarity checks, complex event processing) are targeted and the progress remains impressive. As a result, BPM has a prominent place in today's application landscapes [5]. However, the actual uptake of BPM in the business, as postulated by Hammer and Davenport, did not see the same progression.

Second, and partly as a result of the lack of techniques, methods and systems catering for the ambitions of Hammer and Davenport, organizations which initiated BPR projects often failed dramatically and were not able to replicate the results of the cases as outlined by these two authors [6]. Not unlike other management concepts (e.g., Blue Ocean Strategy), the impression was that outcomes of successful process re-design projects could be observed *ex post*, but there was no reliable way to achieve these. The frequency of BPR failures severely damaged the reputation of management-by-process and put in many cases process improvement projects on hold. Today, many large companies have a BPM Center of Excellence, but it remains typically rather small in scale and impact. In one third of all cases, as our research shows, it will be in the IT department [7].

In light of a methodological and technical landscape of BPM solutions targeting incremental process improvement and automated process execution, BPM as a discipline does not seem to be sufficiently equipped to harvest the potential of an increasingly opportunity-rich environment. One main reason is that current BPM capabilities are largely following an 'inside-out' paradigm, i.e. a process is executed, observable negative deviances and issues are analyzed and addressed where possible. Thus, BPM as it currently stands can be seen as reactive and largely 'opportunity-unaware', i.e. questions such as which of the processes of an organization benefits most from mobile solutions cannot be answered. The significance of this misfit of BPM capabilities is increasing when looking at the substantial changes in the global digital space affording new design possibilities and which have seen the emergence

- of digital public assets with exponential growth attracting user communities of previously unheard scale,
- an ability to outsource infrastructure, data and ultimately processes into the cloud and
- users with fast growing digital literacy possessing under-utilized, mobile, smart assets.

Consequently, it is proposed to complement BPM with an 'outside-in', environmental scanning capability, in which the relevance and impact of external opportunities can be quickly assessed. This will expose business processes to the potential of disruptive innovation and reduce *process innovation latency*, i.e. the time it takes

- to build awareness for the existence of innovation opportunities (data latency),
- to assess the applicability to the internal process landscape and the benefits of this opportunity (analysis latency), and
- to actually implement the opportunity-enabled process innovation (implementation latency).

Figure 1 shows the current, mature inside-out BPM capabilities. However, transformational process innovation rarely results out of the elimination of waste, variation or manual labour along a process. In addition to this inside-out capability, which serves well in an environment striving for predictable, streamlined and efficient processes, organizations aiming for innovation will require complementary outside-in capabilities identifying technological and strategic options and assessing their applicability to the existing or a possible new landscape of processes.

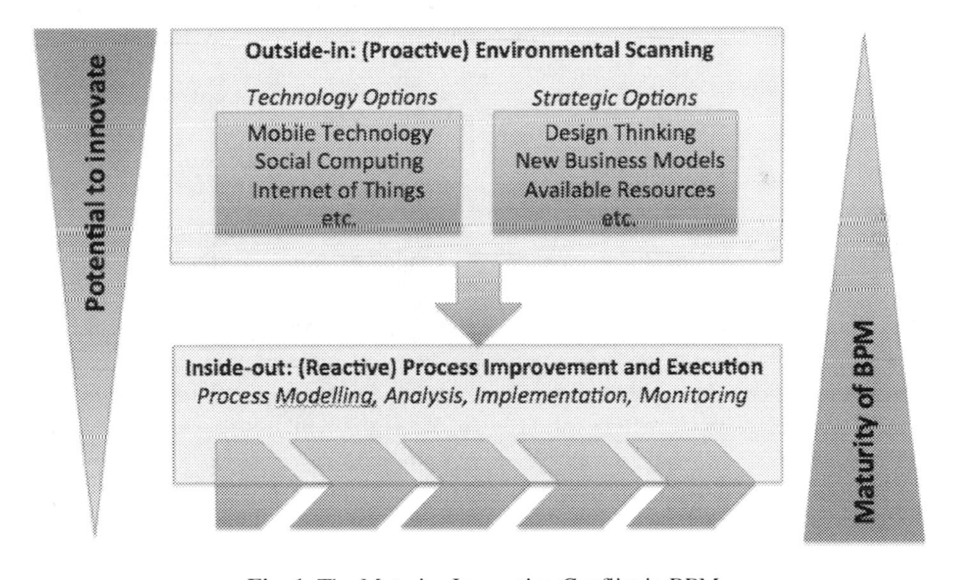

Fig. 1. The Maturity-Innovation Conflict in BPM

In light of this maturity-innovation conflict in BPM, i.e. the misfit between BPM capabilities and an increasingly opportunity-rich environment enabling true innovation, this paper proposes three new research and development directions for BPM. These three directions will be covered in the next chapters before conclusions will summarize the recommendations. First, *Ambidextrous BPM* will be introduced as a new way of conceptualizing BPM consisting of exploitative and explorative BPM. Second, *Value-driven BPM* will postulate a stronger focus on the intended outcomes of a BPM initiative as opposed to the traditional approach of remaining centered on the ability to model and execute processes. Third, *Customer Process Management* will be presented as the ultimate form of outside-in BPM as it puts the customer experiences and their processes at the core of any BPM project.

2 Ambidextrous Business Process Management

2.1 The Ambidextrous Organization

Organizational ambidexterity describes the co-existing corporate abilities of running the current operations as well as being able to continuously adapt the organization to a changing environment. With other words, the ambidextrous organization shows both exploitative and explorative strengths at the same time [8,9].

Exploitation aims towards securing the reliable execution of current business processes. It takes place in the current context of organizational and environmental constraints resulting from strategies, corporate governance, products, services, markets, policies, processes, procedures, regulations, etc. Exploitation secures the ongoing ability to execute business processes according to the promises made to its external and internal stakeholders and in alignment with efficiency expectations, contractual arrangements and compliance requirements. Being unable to *execute-to-promise* can mean ATMs without cash, booking systems unable to take orders, fraudulent currency brokers or failing payroll systems. Depending on the magnitude of the problem, lacking exploitative capabilities can lead even to bankruptcy as failing billing processes in the telecommunication industry have shown. Exploitation includes first loop learning, i.e. a process is monitored, any negative deviants are identified and required adjustments are made. Exploitation is sensitive to the internal process capabilities and an inside-out capability. Consequently, its main metrics are costs, processing time, degree of compliance and further efficiency measures. It is no surprise that techniques, methods and systems supporting exploitation are rather formal, mechanistic, predictable and reliable in nature. Individuals involved in exploitation show high degrees of familiarity with regulations and performance standards and possess strong analytical, engineering abilities. They are mainly interacting with internal stakeholders (business analysts, project managers, various business representatives) and belief in certainty.

Exploration is targeting a much stronger outside-in perspective. Its overall aim is to enable innovation, growth and an effective and efficient capitalization on emerging business and technical opportunities. Unlike exploitation, which is driven by current practices, exploration is focused on future practices, opportunities and risks to the business. Exploration is about imagining a world with driverless cars, always connected customers, location-sensitive services and smart things facilitating new and threatening existing business models and revenue streams. Organizations with strong explorative capabilities follow non-linear, lateral thinking approaches and need to be sensitive to environmental changes in the making. Metrics relevant to exploration are innovation latency and customer-sensitive measures such as new net revenue potential. Individuals tasked with exploration have high sensing abilities, are inspirational and can craft compelling scenarios and visions of the future. They have design capabilities and know how to experiment and prototype. As such, staff tasked with exploration frequently interacts with external stakeholders such as customers, prospects, vendors, analysts or researchers. The world they see is full of uncertainty.

Exploitation and exploration are closely related as exploitative capabilities can be seen as necessary, but not sufficient in a changing environment. An organization not able to even execute-to-promise will have no foundation for far reaching explorative endeavors. This explains why many companies put substantial efforts in building exploitative capabilities before venturing into exploration. Unfortunately, many companies never make it beyond exploitative capability development. The ambidextrous organization, however, is ultimately strong in exploitation *and* exploration. The following table contrasts the views and ambitions of exploitation and exploration.

Table 1. Exploitation versus exploration [10]

Alignment of	Exploitative Business	Explorative Business
Strategic intent	Cost, profit	Innovation, growth
Critical tasks	Operations, efficiency, incremental innovation	Adaptability, new products, breakthrough innovation
Competencies	Operational	Entrepreneurial
Structure	Formal, mechanistic	Adaptive, loose
Controls, rewards	Margins, productivity	Milestones, growth
Culture	Efficiency, low risk, quality, customers	Risk taking, speed, flexibility, experimentation
Leadership roles	Authoritative, top-down	Visionary, involved

2.2 Exploitative BPM

The ideas and principles of the ambidextrous organization can be deployed to the domain of Business Process Management. They allow us to identify those capabilities BPM has to develop to remain of value in the future.

Exploitative Business Process Management is aiming towards running and incrementally improving business processes. It is the best reflection of the current state of the professional and academic BPM community. The existing body of knowledge on how to identify, scope, contextualize, model, analyze, compare, implement, execute, monitor, control and increasingly how to mine and assess the performance of processes can be regarded as being of high maturity.

Exploitative analysis capabilities are dedicated to assessing current processes with the aim to identify and quantify process problems. A large set of exploitative process analysis techniques has been developed and is widely deployed, for example

- *Lean management* with the motivation to locate and eliminate seven types of waste [11]

- *Six Sigma* which, based on a rich set of statistical tools, assesses and reduces the variation of process performance
- The *theory of constraints* targets the elimination of bottlenecks in the process and comes with a set of guidelines for how to overcome such bottlenecks [12]
- *Process modeling* can be seen as an approach targeting the lack of shared understanding among process stakeholders
- *Workflow management* and *straight through processing* approaches aim towards the replacement of manual labor via automation along business processes

Further process analysis techniques such as SIPOC (what is the context of the process?), viewpoint analysis (what process parts are visible to the accounting department?), scenarios analysis (how do loan applications below $1mio flow through the system?), Pareto analysis (do 20% of all processes explain 80% of all issues?) or process simulation (how does the process behave under different loads or with alternative resourcing?) increase the overall transparency in the analysis. None of these, however, is able to generate actual improvement ideas as an outcome.

Exploitative execution capabilities are dedicated to the reliable, automation of business processes taking into account the varying requirements of different types of business processes. Available solutions include workflow management, service-oriented applications, exception handling, (adaptive) case management or document management, but also hard-coded processes as they can be found in Enterprise Systems or industry-specific solutions (e.g., banking, insurance, higher education).

Exploitative BPM serves well in industries and organizations with largely static market conditions (e.g., banking back-offices, shared service providers, mass production). The efficient execution of processes secures economies of scale, leads to high levels of predictability, control and transparency. It simplifies resourcing, costing and overall planning decisions. Historical data as derived from event files has a high value and allows informing future process design activities.

Exploitative BPM capabilities in the form of methods and systems are widely available on the market and often obtainable for free (e.g., process modeling editors). Exploitative analysis and execution techniques have made it into the curriculum of many universities and more recent generations of business analysts are nowadays well equipped for all BPM challenges related to execution-to-promise. In many cases, exploitative BPM can be even regarded as a commodity, i.e. it is a corporate expectation that processes can be modeled, analyzed, implemented and executed. The successful execution of exploitative BPM, unless substantial operational gains are the result, hardly leads to excitement in the boardroom anymore. BPM centers of excellence within organizations that are purely focused on exploitative BPM tend to remain rather small in size and have limited, enterprise-wide visibility and impact.

If exploitative BPM can be indeed regarded as a commodity, a hygiene factor, it is not without its own challenges. Under-utilized process model repositories, process analysis projects consuming substantial resources and lasting for many months or failures in the execution of processes lead to substantial criticisms. In corporate capabilities that have become a commodity, the involved stakeholders will hardly ever receive the recognition they desire.

The last two decades have seen the growth of substantial exploitative BPM capabilities, and these achievements need to be applauded and form an excellent foundation. However, I recommend moving towards higher aspirations and channel future development and research efforts from exploitative to explorative BPM. What is needed in an opportunity-rich environment are revenue-sensitive BPM approaches facilitating the design of entire new process experiences capitalizing on emerging technical solutions and satisfying a consumer base with increased digital literacy.

2.3 Explorative BPM

A pure exploitative approach to business processes will not be sufficient for the future as an exclusively internal, reactive focus on streamlining existing business processes takes the eye of the customer of a process. This is evidenced by the fact that hardly any company conducting BPM initiatives involves its customers (or prospects) into the design of its future processes.

The shortcomings of exploitative BPM are particular dramatic in industries exposed to the disruption of digital innovation meaning in many cases the emergence of an entire new class of competitors. The most famous, recent example is Kodak, an organization that at its peak employed 140,000 people, had a market value of $28billion and invented in 1975 the digital camera (!). The disruption of digital photography as showcased by Instagram was one factor that took Kodak despite all its internal exploitative BPM capabilities out of business. Kodak remained focused on products and processes further materializing its belief of being in the film industry. The fact that Instagram was sold for $1billion to Facebook and at that stage had 13 employees shows the diminishing role of scale. Further examples in the making are retail banks (competition: social media) and logistics and medical device providers (3D printing). The following more current examples show that a cost-effective process does not secure ongoing revenue streams.

- Over-the-top applications such as WhatsApp or WeChat have started to eliminate SMS-related revenue from telecommunication providers
- The peer-to-peer platform Airbnb has become a severe threat for the hotel industry
- Paypal has extracted substantial revenue from retail banks
- CourseERA and EdX are recent examples for how the provision of massive open online courses poses a threat to the revenue model of universities
- Services such as Uber and Lyft have started to cannibalise the market share of established taxi companies

Explorative BPM is a significant future opportunity, and challenge, for the BPM community. The techniques, methods and systems required here need to substantially go beyond what is available at this stage. The following points outline a few of the desired capabilities of explorative BPM. At its core is the facilitation, and where possible the (semi-)automated derivation of new processes.

Explorative BPM is about crafting *process visions* that are so compelling and transformational that they motivate staff, and customers, involved to explore (!) how to make a desired future state via a sequence of transition states a reality, and by this the current process obsolete. This is in sharp contrast to exploitative BPM, which develops new (to-be) processes in light of current shortcomings. The idea of compelling process visions goes back to Charles F. Kettering, research chief at General Motors who in the early 1920s did not want to accept that painting a car (by hand) needed to take 37 days [13]. While Kettering's engineers believed it would be possible to reduce the processing time to 30 days, his vision of bringing this part of the car manufacturing process down to an hour led to the search for entire new opportunities far outside the immediate vicinity of the process. Kettering found the solution in the form of a new lacquer at a jewelry store in Manhattan. With the help of DuPont, a liquid was engineered that could be spray-painted and dried in minutes.

Current BPM techniques are not able to generate such process visions. Instead of the dominating, incremental reductionist approach (e.g., eliminate waste, variation, bottlenecks or manual work), entire new methods are required allowing to design such compelling scenarios and ways to achieve these. Further examples for such process visions are a bank aiming for a 24hrs mortgaging process, Amazon's delivery via drones, its anticipatory shipping concept (*'we know before you know what you will buy'*) or a federal government aiming to renew a passport before the citizen notices that it expired. Many of these visions are inspired by current technologies and an ability to transfer their affordances into new process design opportunities. They are a result of seeing the capabilities behind technologies. BPM explorers have to have very ambitious goals and must be able to strip a process back to its most basic core.

Whereas exploitative BPM is centered on the construct of 'pain points' within a particular process, explorative BPM is about the identification of *opportunity points* in processes. Such opportunity points capture where in the collection of its processes a retailer will benefit from facial recognition, where an insurer could utilize location-based services, where a public sector agency could offer citizen-to-citizen brokerage services or where a travel agency could capture external, social signals for more proactive customer interactions. Such opportunity points need to be conceptualized with precise semantics and contextualized in existing BPM techniques, methods and systems. Ways for how to capture these within a process, but also as part of a process modeling query language are to be developed.

The environmental scanning that comes with explorative BPM can be supported by a trading place for truly inspirational, exciting business processes. Unlike the first wave of reference models, which have been developed for industries and disciplines such as telecommunication (eTOM), IT service management (ITIL), supply chain management (SCOR) or enterprise systems (e.g., SAP), future reference models need to go substantially beyond such common sense models. These large collections of reference models provided important foundations for organizations about to engage in BPM in the large. However, these models rarely have been a source of breakthrough innovation. In the spirit of *open process innovation*, new exciting reference process models will have to be much smaller processes or process parts. They would need to have a short latency, i.e. emerging technologies need to quickly be converted into such models (e.g., how do conduct pay-as-you-drive insurance services; collaborative consumption opportunities in online retail). The BPM community could play an important role in proposing such opportunities via formally defined and accessible reference processes. Assuming the right meta-tags are available, organizations could even subscribe to such a process innovation marketplace and would be notified, if innovations of potential relevance have been made available.

A further idea for developers and researchers committed to boosting explorative BPM could be the development of *process improvement/innovation systems* (PIS) providing services to existing BPM tools. The user of the future would be able to highlight parts of the process landscape and such a PIS would semi-automatically propose possible process designs of interest using artificial intelligence in the form of fuzzy logic, machine-based learning or case-based reasoning. A possible scenario could be that a process analyst investigates a process including an invoicing activity. A PIS would among others propose the solution 'Usage-based pricing' and suggest to move the point of invoicing behind the point of consumption. Such a solution, while popular in car parks and for phone companies, might be of high interest for shared service providers, logistical service companies or tourist attractions. The role of the PIS is to rapidly increase the accessible, relevant solution space for the analyst. Developing a rich set of such process improvement patterns and related *process recommender systems* is a path of high relevance, but widely unexplored.

3 Value-Driven Business Process Management

Current BPM capabilities are centered on developing methods, tools and systems, and less about actual processes. This is noticeable in the nature of BPM papers and in the presentations from academia and BPM professionals. They largely report on the procedural aspects of the process of process management, and describe intermediary outcomes such as process architectures, process modeling techniques or BPM offices. However, in many cases they fall short in terms of reporting BPM's actual achievements. A discipline that is more focused on how it conducts its work rather than gathering evidence for the existence of its value propositions faces compromises to its credibility by those who take a black-box view on it.

Therefore, in a joint research project with Accenture [14], we wanted

(1) to identify those values that truly matter for BPM initiatives, and

(2) assess the extent to which BPM solutions support each of these values.

Grounded in a comprehensive literature review, this research involved a series of global focus groups in London, Sydney and Philadelphia with selected clients of Accenture. Our research outcomes showed that the frequent lack of an explicit focus on the intended outcomes of BPM is a main reason for its limited credibility. In addition, the dominant activity-driven, internal nature of BPM initiatives means that these projects rarely on the critical path of corporate development. Moreover, even if values are identified that drive a BPM initiative, existing BPM methodologies can often not be tailored to these specific values.

Value-driven BPM (VBPM) extends the current body of BPM knowledge and practices by giving priority to the objectives that drive a BPM initiative. Rather than following traditional BPM practices and concentrating on mapping the organization in hierarchies of value chains, VBPM starts with the "Value-Value-Chain, i.e. what needs to be done to achieve the outcomes which motivated the BPM project in the first place. It raises issues such as how BPM can contribute to the strategic agenda of an organization, how to make processes tangible and help to overcome classical business conflicts. VBPM means process management that can be tailored to the values that trigger the BPM initiative. Our research shows that organizations aim towards different values when starting a BPM initiative. These values can be summarized as one core value and three value pairs.

Transparency is at the core of the VBPM framework, and is fundamental to achieving any of the other six values. Only an organization that has a shared understanding of its processes can start reflecting on better ways to design and operate them. Thus, transparency is a necessary condition for VBPM. Research on tangible process modeling [15] or the use of virtual environments [16] are attempts to increase the transparency of processes and the ease-of-engagement in process design activities.

The six values can be grouped into three value pairs. While each of these pairs consists of two values that tend to be oppositional, BPM has the potential to moderate and ease these traditional conflicts.

The *efficiency-quality pair* reflects the widely accepted dichotomy of Porter's strategic core alternatives [17], i.e. a focus on streamlined, highly productive operations or a concentration on a customer-focused, quality-driven strategy.

The *agility-compliance pair* depicts the requirement to be highly adaptive and flexible versus the increased demand to ensure that operations are conducted predictably and according to compliance standards.

Finally, the *integration–networking pair* captures the fact that organizations can concentrate on integrating their employees in the design of processes or focus on networking with and benefiting from the input of external partners and resources. These three pairs are not strictly oppositional, and many organizations will actually have to address all six of these values in some form during their BPM initiative. However, our research shows that a BPM initiative can be characterized by choosing priorities within this value framework and within each of the three value pairs.

Three of these values capture internal goals, including efficiency, employee integration and compliance. In contrast, the other three values - quality, agility and networking - reflect values with an external focus (Figure 2).

In order to increase the value-sensitivity of BPM, researchers are encouraged to select values of BPM and start customizing the current set of BPM capabilities. This would extend the fast growing body of knowledge on configurable processes to the domain of *configurable Business Process Management.*

Also labeled *X-aware BPM*, value-driven BPM requires tailoring and expanding BPM to the specific demands of a value. For example, BPM centered on the integration of internal employees needs to address questions as how to model, capture and increase employees' satisfaction with the business process and the activities involved. One possible path to explore would be, if the inclusion of a *'Like it'* button allowing staff to express an interest in such activities could extend existing work allocation principles towards an increased preference-based allocation of work within workflows.

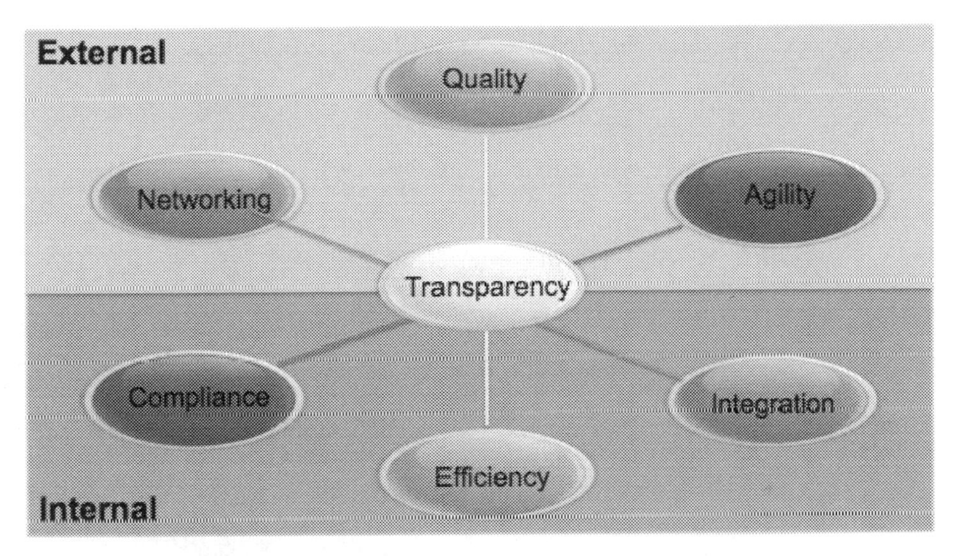

Fig. 2. The Values of Business Process Management [14]

Further relevance for Value-driven Business Process Management will be in the fast growing world of '*big process data*'. Data science and process mining are fields of interest and provide a rich set of analytical capabilities. However, without a close link to decision science and a sound understanding of the value such big data sets are supposed to generate, this community lacks a direction. An example can be seen in an approach to process mining following classical BPM values such as cost or time efficiency. In this case, a process mining exercise might identify negative deviants and trigger reactive process improvement activities aiming to overcome this issue. A focus on alternative values, e.g. revenue or customer satisfaction, combined with correlating event files with further case data can channel such process mining initiatives towards the positive deviants, i.e. identifying the future to-be processes within the as-is processes. In this case, the task will be to identify the distinct process patterns of such positive deviants and to explore the extent to which these patterns can be replicated by other stakeholders in a similar environment. Our own experiences in the domain of fresh-food retail and cross-organizational insurance processes have pointed to substantial gains in such process mining activities resulting from a simple shift in the value driver of BPM.

X-aware BPM could go far beyond the identified values and include strategy-aware, culture-aware, risk-aware, resource-aware, knowledge-aware, location-aware, context-aware, emission-aware or data-ware BPM. Some of these would not target immediate business values, but intermediary steps (e.g., context-aware BPM is one way to decrease corporate latency,and emission-aware BPM is one way to reduce the environmental footprint of an organization). Researchers will be required to build up deep expertise of the relevant value (e.g., what are the exact requirements of risk management?) and then develop appropriate solutions. This shows how value-driven BPM will go beyond the core goal of BPM, i.e. the reliable execution of processes with a focus on processing-time. It will in some cases also act as a counter balance to the paradigm of one-dimensional process optimization. For example, resource-aware BPM will need to balance the conflicting targets of managing idle time and waiting time and hardly ever lead to the process with the shortest time. Thus, value-driven BPM will ensure that BPM efforts are contextualized in light of a (global) corporate optimum, not just local process objectives.

4 Customer Process Management

Today, Business Process Management is focused on internal business processes such as procurement, manufacturing, sales or payroll. Despite the fact that many organizations claim to take a 'customer-centered' view on the process, actual customers are hardly ever involved in the analysis or design of business processes. Even further, the customers' process experiences before they consume the provider-specific business processes are often not captured. For example, the end-to-end process scope of a financial service institution providing mortgage services is a very small subset of the end-to-end experiences of a customer buying a house.

Customer Process Management is strongly aligned with the claim for a stronger outside-in focus on BPM. It is ultimately grounded in the existence of a *birth-to-death value chain*, i.e. all business processes are directly or indirectly derived from the value chain of the life of a customer.

CPM as opposed to BPM demands a mind-shift as beautifully articulated by Chris Potts [18]: *It is not about how customers participate in our (business) processes, but about how we participate in the customers processes."*

Following this view means processes start way before the customer contacts an organization, and are triggered by life events experienced by the customer (such as a wedding that triggers legislative processes to change names). An example for a customer process would be the car manufacturer who calls the driver when it is noted that the car had an accident. Failing a response, the manufacturer might consider sending an ambulance. In a similar way, health care solutions are now capable of sensing personal health data and can trigger actions when needed.

The affordances of the Internet of Things [19] and social media have strengthened the role of events, the 'poor cousin of BPM'. The dominating focus of BPM has been on activities and their cost-effective orchestration considering time and quality requirements. Events have played a role as part of architectures and models, but where typically not a first class citizen in business conversations regarding processes. In the world of customer processes, however, *process execution latency*, i.e. the time it takes a process to detect a possibly relevant event, analyze its relevance and create a process instance, if needed, has become a source of competitive advantage. Complex-event processing has started to explore this domain, but there is much to research here including how to design trusted processes ensuring unconditional privacy.

The more customer processes will be designed and supported, the more corporations will observe a shift from their business processes as the mode of engagement for their external stakeholders to the requirement of a seamless participation in the processes of their increasingly digitally enabled and literate customers. For example, when the Australian airline Qantas offered RFID-enabled frequent flyer cards to their passengers, it eliminated the demand for boarding passes. A boarding pass is a typical artifact that is necessary for a customer to participate in the business process of an organization. A customer would not request a boarding pass. Boarding by simply using a frequent flyer card, or fingerprints like in Sweden, means taking part in the processes of the customer. The interaction is reduced to a request for identification, not the existence of an artifact required by the company.

For the BPM community to engage with the idea of Customer Process Management, it needs to develop stronger design capabilities covering empathy, integrative thinking, optimism, experimentalism and collaboration [20]. An emerging *Design-led Process Innovation* stream has the potential to replicate the success organizations had with the design of customer-centered products and services to the domain of processes. This will facilitate the identification of often hidden customer expectations in existing processes and open up entire new insights into process experiences that start far before the corporation is engaged.

5 Conclusions

The members of the BPM community are as human as everybody else. We are creatures of habits, look for certainty and exploit available skills and expertise. This is one explanation why we today observe a very high level of exploitative BPM capability, both in practice and in research. BPM tools, methods and systems still offer countless challenges requiring the development of new algorithms and solutions, and practitioners will continue to struggle to develop good process models and finding the true root causes of an as-is processes. However, the allocation of substantial resources, especially in research, into process modeling, process analysis, process execution and process mining has also taken away the focus on more disruptive but also less predictable process innovation challenges. These are in high demand when fast emerging technological opportunities need to be translated into value-adding affordances for corporations and their customers. This demand will hopefully encourage new talent from neighboring disciplines to join the BPM community.

This paper proposed under the label of 'Ambidextrous BPM' extending the strong exploitative BPM capabilities via future BPM research and development into the domain of explorative BPM. This direction provides countless opportunities to study how BPM solutions of the future can provide more advanced and proactive support in the quest for better processes.

Value-driven Business Process Management has been tabled as a way to re-sensitize the BPM community for the importance of the actual outcomes of BPM, i.e. tangible contributions to corporate goals. Researchers are invited to consider establishing a body of knowledge on configurable BPM, which will lead to BPM tools, methods and systems catering for the specific needs of individual BPM initiatives. This will in most cases require cross-disciplinary efforts in order to comprehend the specific needs of, for example, cost accounting, risk management or sustainability consideration.

Customer Process Management (CPM) is the ultimate form of an outside-in view on BPM. Building a Design-led Process Innovation capability will allow crafting processes tailored to the desired experiences of customers. Going beyond individual providers, might even lead to semi-automated customer processes in which not the customer, as at the moment, but CPM solutions will take over parts of the orchestration. The demand for such solutions will increase with the extent to which such processes are regulated.

No matter what the future will hold, BPM researchers have plenty of opportunities to explore new avenues. Like most process changes, this might, however, require a re-adjustment of current activities and true research into the yet unknown.

References

1. Hammer, M.: Reengineering Work: Don't Automate, Obliterate. Harvard Business Review, 104–112 (1990)
2. Davenport, T.: Process Innovation: Reengineering Work Through Information Technology. Harvard Business Review Press (1992)
3. Taylor, F.: The Principles of Scientific Management. Harper & Brothers (1911)
4. Scheer, A.-W.: ARIS - Business Process Modeling, 3rd edn. Springer (2000)
5. van der Aalst, W.M.P.: Business Process Management. A Comprehensive Survey. ISRN Software Engineering, Article ID 507984 (2013)
6. Sarker, S., Lee, A.S.: IT-enabled organizational transformation: A case study of BPM failure at TELECO. Journal of Strategic Information Systems 8, 83–103 (1999)
7. Rosemann, M., Lehmann, S.: zur Muehlen, M., Laengle, S.: BPM Governance in Practice. Accenture: Philadelphia (2013)
8. Duncan, R.: The ambidextrous organization: Designing dual structures for innovation. In: Killman, R.H., Pondy, L.R., Sleven, D. (eds.) The Management of Organization, pp. 167–188. North Holland, New York (1976)
9. March, J.G.: Exploration and exploitation in organizational learning. Organization Science 2, 71–87 (1991)
10. O'Reilly, C.A., Tushman, M.L.: The Ambidextrous Organization. Harvard Business Review, 74-81 (2004)
11. Womack, J.P., Jones, D.T.: The machine that changed the world. Free Press (2007)
12. Goldratt, M.E.: Theory of Constraints. North River Press (1999)
13. Leslie, S.W.: Boss Kettering: Wizard of General Motors. Columbia University Press (1983)
14. Franz, P., Kirchmer, M., Rosemann, M.: Value-driven Business Process Management, Philadelphia, USA (2011)
15. Edelman, J., Grosskopf, A., Weske, M.: Tangible Business Process Modeling: A New Approach. In: Proceedings of the 17th International Conference on Engineering Design. Stanford University, Stanford (2009)
16. Brown, R., Recker, J., West, S.: Using virtual worlds for collaborative business process modeling. Journal of Business Process Management 17(3), 546–564 (2011)

17. Porter, M.: Competitive Strategy. Free Press (1980)
18. Potts, C.: RecrEAtion: Realizing the Extraordinary Contribution of your Enterprise Architects. Technics Publications (2010)
19. Rosemann, M.: The Internet of Things – New Digital Capital in the Hand of Customers. Business Transformation Journal 9, 6–14 (2014)
20. Brown, T.: Design Thinking. Harvard Business Review, 84-92 (2008)

A Universal Significant Reference Model Set for Process Mining Evaluation Framework

Qinlong Guo[1], Lijie Wen[1], Jianmin Wang[1], Zizhe Ding[2], and Cheng Lv[1]

[1] School of Software, Tsinghua University, Beijing 100084, China
[2] Department of Management Information System,
China Mobile Communication Corporation, Beijing 100033, China
guoqinlong@gmail.com, wenlj00@mails.tsinghua.edu.cn,
jimwang@mail.tsinghua.edu.cn, dingzizhe@chinamobile.com, 425535361@qq.com

Abstract. Process mining has caught the attention of researchers and practioners. Because a wide variety of process mining techniques have been proposed, it is difficult to choose a suitable process mining algorithm for a given enterprise or application domain. Model rediscoverability of process mining algorithms has been proposed as a benchmark to address this issue. Given a process model (we call it original model) and its corresponding event log, the model rediscoverability is to measure how similar between the original model and the process model mined by the process mining algorithm. As evaluating available process mining algorithms against a large set of business process models is computationally expensive, some recent works have been done to accelerate the evaluation by only evaluating a portion of process models (the so-called reference models) and recommending the others via a regression model. The effect of the recommendation is highly dependent on the quality of the reference models. Nevertheless, choosing the significant reference models from a given model set is also time-consuming and ineffective. This paper generalizes a universal significant reference model set. Furthermore, this paper also proposes a selection of process model features to increase the accuracy of recommending process mining algorithm. Experiments using artificial and real-life datasets show that our proposed reference model set and selected features are practical and outperform the traditional ones.

Keywords: Universal Significant Reference Model, Process Mining, Feature Selection.

1 Introduction

Process mining is a process management technique that learns process models from event logs. The mined models will then be compared against the original process models of the enterprise for conformance checking, or for discovering more efficient, streamlined business process models. Today, there exist a wide variety of process mining algorithms. However, since different process mining algorithms have different characteristics (e.g., some perform better on models

C. Ouyang and J.-Y. Jung (Eds.): AP-BPM 2014, LNBIP 181, pp. 16–30, 2014.

with non-free choice, while others do not), and different enterprises' models have different features (e.g., process models of a train manufactory are much different from those of a boiler factory), being able to select the most appropriate process mining algorithm for the models from a given enterprise is a big advantage.

Nowadays, model rediscoverability [3] has been proposed and accepted as a criterion on choosing suitable process mining algorithm for a given model set. Given a model and its corresponding event log, model rediscoverability of a process mining algorithm is measured by the similarity degree between the original model and the model mined by the process mining algorithm.

In real-life scenarios for enterprises, event logs are usually recorded by information systems, and indicate how enterprises' processes work in practice. While process models are usually pre-designed, and shed light on how enterprises' processes are supposed to be. Therefore, model rediscoverability expresses the degree enterprises' processes work as expected, and this is an important concern of enterprises. However, model rediscoverability has a serious limitation that empirically evaluating all available process mining algorithms against a large model set (e.g. from a large group company) can be tedious and time-consuming.

Some recent works (e.g. [1], [4]) have been done to accelerate the evaluation of model rediscoverability by empirically evaluating a small set of process models (called the reference models) and recommending the others via a regression model. The performance of recommendation is highly dependent on the reference models. The quality of reference models is measured by the degree how process models differentiate the process mining algorithms. The models that lack the capability of differentiating process mining algorithms are not suitable as the reference models. For example, the sequential models in Figure 2 can be rediscovered correctly by almost all process mining algorithms, and they are fruitless as reference models. The process models which are good at differentiating process mining algorithms are called *significant reference models*.

To the best of our knowledge, the only approach dealing with the selection of significant reference model set is proposed in [1]. This approach is a training-classification procedure. Firstly, a portion of process models from a given model set are randomly selected as training set. Then, from the training set, the approach obtains a classification model which tells whether a process model is a significant reference model or not. At last, other reference models are determined based on the classification model.

However, this approach has several limitations. First, it does not completely avoid empirically evaluating process models. Because reference models are selected from the given datasets, significant reference models are different in different datasets. For each dataset, in order to obtain the regression model, we have to empirically evaluate its own reference models. After all, though this approach does not need to empirically evaluate the whole dataset, it has to evaluate the reference dataset. Moreover, this approach cannot guarantee the selected reference models are all significant. Because this approach is based on probability, it is possible to contain some unsuitable process models in the reference models.

In order to overcome the limitations mentioned above, the primary aim of this paper is to generalize a universal significant reference model set. This model set can be trained once, then the training result can be used without extra training for each different reference model set. Accordingly, with this model set, it can save lots of time to select and evaluate reference model set from each different dataset. Besides, the quality of universal significant reference model set is guaranteed to be significant, since these process models are selected by our reasonable analysis, rather than a probability-based method.

Several recent works have been done on depicting the model features ([2], [1]). In [1], a set of process model features (totally 48), which are extended from [5], are used to obtain the regression model. These model features measure process models in a wide variety of aspects. However, model rediscoverability is the main concern in recommending process mining algorithms, and lots of model features used in [1] are unrelated to this aspect. These unrelated features may have a bad influence on the accuracy of the regress model, and increase the time cost of extracting the features. In this paper, we inspect and analyze these process model features, remove unrelated features, and reduce the size of metrics from 48 to 6.

We managed to conduct experiments using both artificial and enterprise datasets. Experiments results show that our selected model features and the proposed universal significant reference model set outperform the traditional approach in [1].

In summary, there are two main contributions for process mining evaluation framework in this paper:

1. A universal significant reference model set is proposed.
2. A small set of process model features that are specializing on model rediscoverability are selected.

The organization for the rest of this paper is as follows. Section 2 summarizes the overview framework of efficiently evaluating process mining algorithm with reference models (we call it *EVA* framework for short). Section 3 discusses selecting the features that depict the model rediscoverability. Section 4 describes the universal significant reference model set. Section 5 presents the experiments and discusses the result. Eventually, Section 6 concludes the paper, and shows the future work.

2 Efficiently Evaluating Process Mining Algorithm with Reference Model Set

The *EVA* framework is built by extending the work from [6]. Figure 2, which is first proposed in [1], shows the overall architecture of this framework. This framework lay the foundation for our work in this paper.

The original idea from [6] takes the original reference models or the process logs as the input. The process mining algorithms are then run on these logs to

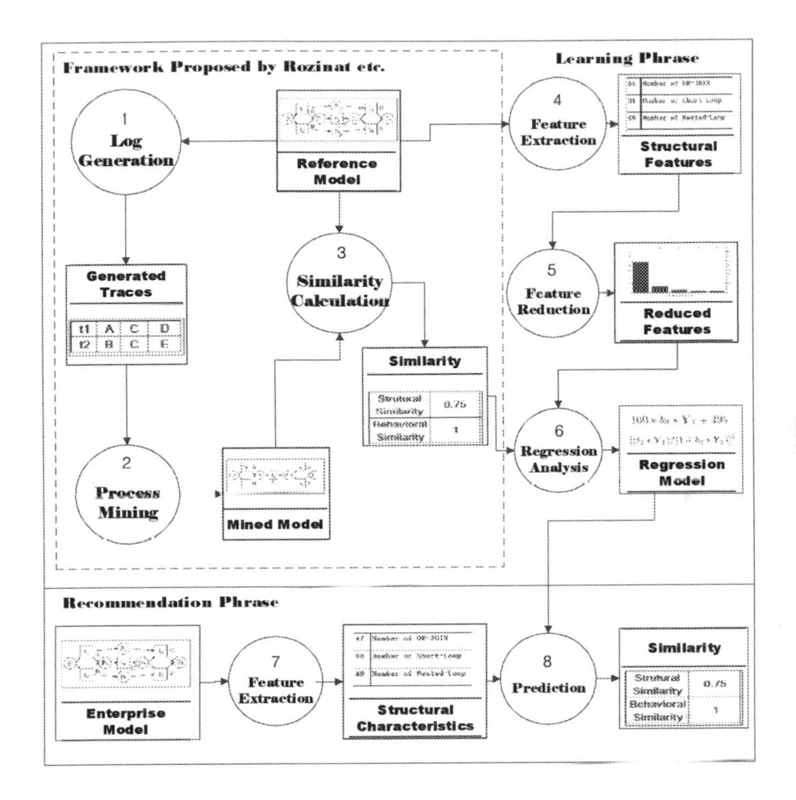

Fig. 1. Overall architecture of *EVA* framework

produce the mined models. Next, the similarity measure can then be performed on the mined models and the original models.

As shown in Figure 2, the *EVA* framework includes a *learning phase* and a *recommendation phase.*

During the *learning phase*, the reference models are used as training set to obtain the regression model. The feature extraction module extracts features based on the model metrics, in order to build a regression model that captures the similarity measure. Since the features extracted cover a wide variety aspects of models, an unnecessary number of features may be extracted in this module. Feature reduction (the PCA techniques) is used to reduce the features. Finally, during the learning phase, regression analysis produces a regression model that captures the correlation between the features and the similarity values.

After the regression model is obtained, during *the recommendation phase*, the most suitable (i.e. best performing) mining algorithm can be recommended without performing the actual empirical benchmarking. Since applying the regression model to the model features to obtain the estimated similarities is very fast, the only cost during the recommendation phase is feature extraction.

While various similarity measurements for business process models have been proposed recently, e.g.,[16,17], in this paper, we adopt the Principle Transition Sequence (PTS) metric [10] as the similarity metric, since it can deal with many complex structures, such as non-free choice, arbitrary loop, even the unbounded Petri net, etc, and can be calculated efficiently. However, please note that the *EVA* framework can easily incorporate other measurement scheme to meet any specific requirement.

The process mining algorithm being evaluated in this paper are the α algorithm [7], α^{++} algorithm [11], $\alpha^{\#}$ algorithm [12], the genetic algorithm [13], the duplicate task genetic algorithm [13], the heuristics miner [14], and the region miner [15]. Without loss of generality, we believe that these seven algorithms represent the several types of mainstream business process mining algorithms. Besides, akin to the similarity algorithms, other process mining algorithms can easily be integrated into this framework.

With the universal significant reference model set and the selected 6 process model features, this framework has three main improvements.

1. The reference models are the universal reference models rather than the process models selected from each dataset. This change makes the module 1-6 unrequired for each dataset. Without the universal significant reference model set, the reference models are the enterprise models coming from each dataset, thus the model training (i.e. module 1-6) has to be executed for each different dataset. Since the universal reference model set is independent with datasets, training can be done only once, and the training result (i.e. the regression model) are used without extra training for each different dataset.
2. The features extracted in the *Feature Extraction* module (the module 4 and 7) are the selected 6 features rather than the 48 features. These features perform better than the whole 48 features with respect to the model rediscoverability.
3. The *Feature Reduction* module 5 is excluded. Since the size of the features extracted in the *Feature Extraction* is largely reduced, and these selected features focus on depicting the model rediscoverability, the *Feature Reduction* module is not useful any more.

3 Feature Selection

In [1], totally 48 features are proposed and applied to depict a process model. Many of these process model features are not related to the model rediscoverability, such as the *number of transitions*, *number of xor-split*, and so on. These irrelevant process model features may have bad influence on the recommendation. Besides, it needs extra time to extract these unnecessary features. Therefore, we inspect and analyze these 48 features, and rule out the unnecessary ones.

We propose two criteria for selecting the suitable process model features regarding model rediscoverability. Besides, note that the process logs of models are generated automatically in our framework, they are guaranteed to be complete without noise.

1. The features which characterize a model's size should be removed.
2. The features which characterize a model's connectors should be removed.

Features like *number of nodes* and *diameter* meet the first criterion. A plain example to illustrate this criterion is shown in Figure 2. Both models are simple sequential petri nets, with remarkably different sizes (one has 2 transitions, the other has 101 transitions). Although these two models have significantly different *number of nodes* and *diameter*, both of them can precisely be rediscovered by almost all process mining algorithms.

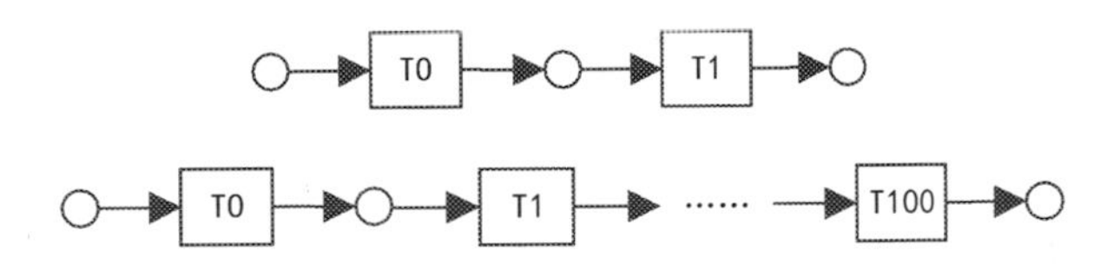

Fig. 2. Two sequential process models with significantly different size

Connectors are the elements represent the router information in a process model. For example, when considering the Petri Net, the connectors include the AND-split, AND-join, XOR-split, XOR-join. Features meeting the second criterion include *Number of connectors* , *Connector heterogeneity*, *Control-flow complexity*, etc. The reason for this criterion is that [8] points out the structured workflow net can be properly rediscovered by almost all process mining algorithms, and the connectors (e.g. AND-split, AND-join) are supported in the structured workflow net. For instance, two process models in the Figure 3 are sharply different in connectors(e.g. the above one has a xor-split connector with 3 output edges, while the below one has a xor-split connector and an and-split.). However, both of these two process models are structured workflow net, and can be mined properly by most process mining algorithms.

Based on these two criteria, 6 suitable features are selected at last:

1. *Number of invisible task* : the number of invisible tasks in a model.
2. *Number of duplicate task* : the number of duplicate tasks in a model.
3. *Number of non-free choice* : the number of non-free choices in a model.
4. *Number of arbitrary loop* : the number of arbitrary loops in a model.
5. *Number of short loop* : the number of short loops in a model.
6. *Number of nested loop* : the number of nested loops in a model.

4 The Universal Significant Reference Model Set

We have collected 60 process models as the universal significant reference model set. Inspired by the selected 6 process model features that characterize the model rediscoverability, we artificially construct 10 process models for each process model feature, as the universal significant reference model set.

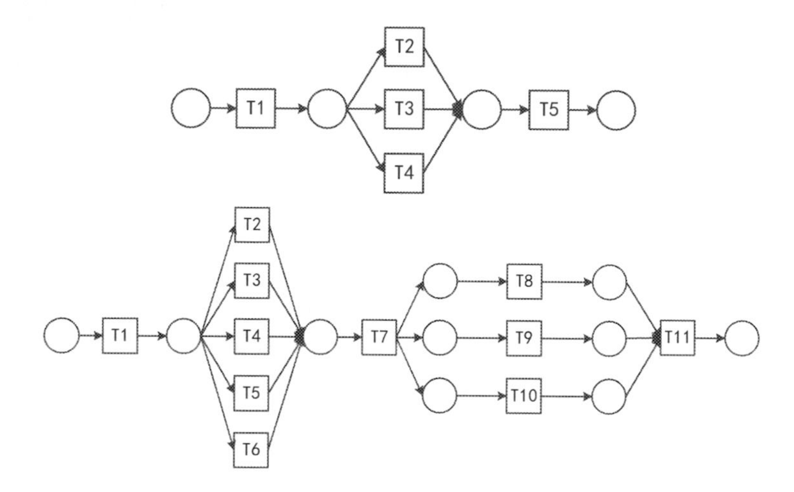

Fig. 3. Two sequential process models with significantly different connectors

Figure 4(a)–4(f) respectively show basic properties of these 6 subset. We measure the *number of transitions*, *number of places* and *number of edges* for all subsets. Furthermore, for each subset, we separately measure its own corresponding feature. For instance, we measure the *number of invisible task* for the invisible task subset.

4.1 Invisible Task

Invisible tasks are such tasks that exist in a process model but not in its event log[12]. As invisible tasks do not appear in the event logs, they increase the difficulty of mining the proper process model. For example, Figure 5 is a process model with an invisible task. Since the invisible task is not registered in the event log, task T6 is directly after task T5 in some trace of the event log.

4.2 Duplicate Task

Duplicate task means that a process model can have two or more tasks with the same label. A process model with duplicate tasks is shown in Figure 6. In this model, there are two tasks referring to T4. However, it is difficult to construct this process model from its corresponding event log, because it is hard to distinguish the tasks referring to the same label.

4.3 Non-free Choice

A Petri Net with non-free choice is such a petri net where there exists at least one input place shared by a pair of transitions, but the input places of these transitions are not the same. For example, Figure 7 is a process model with a

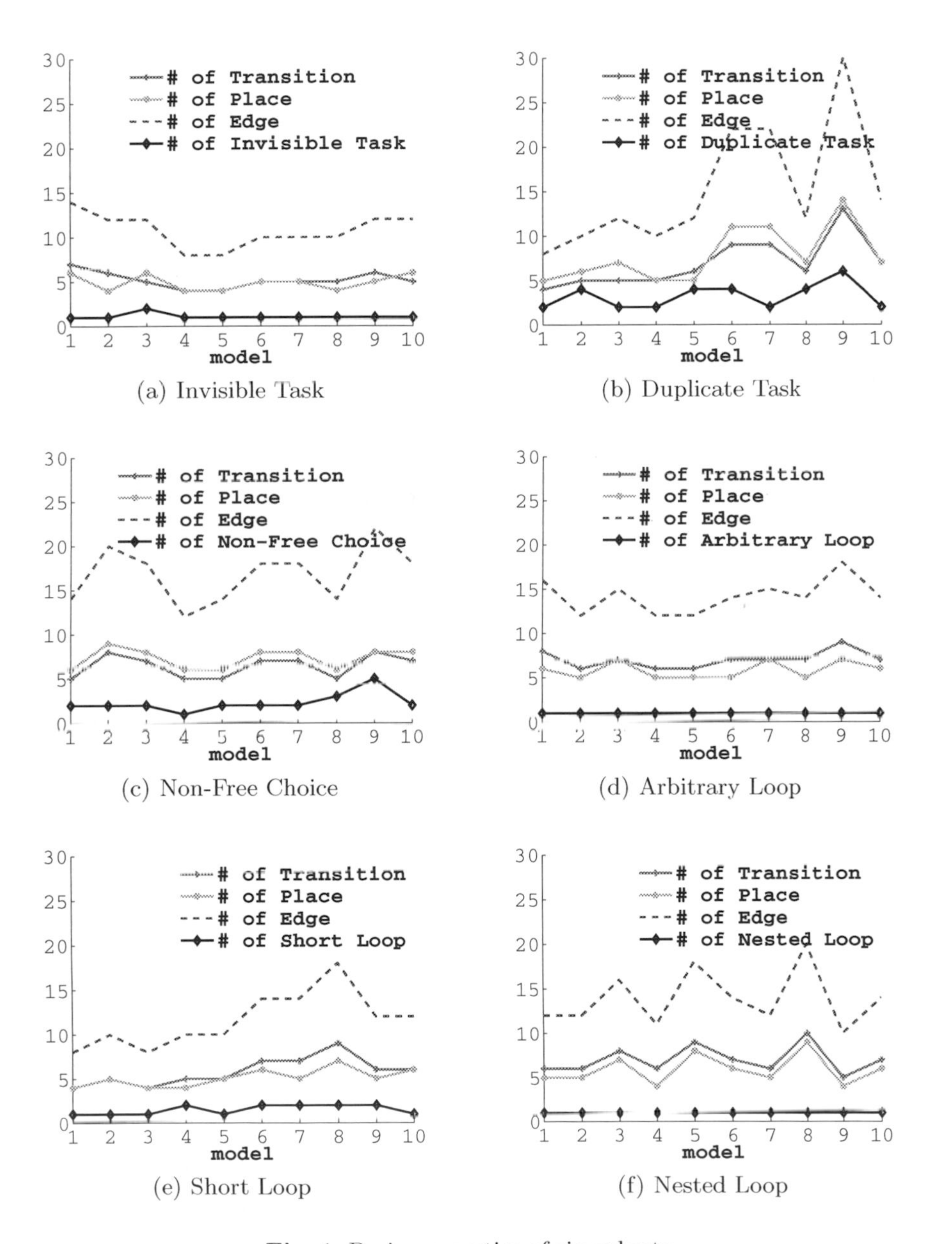

Fig. 4. Basic properties of six subsets

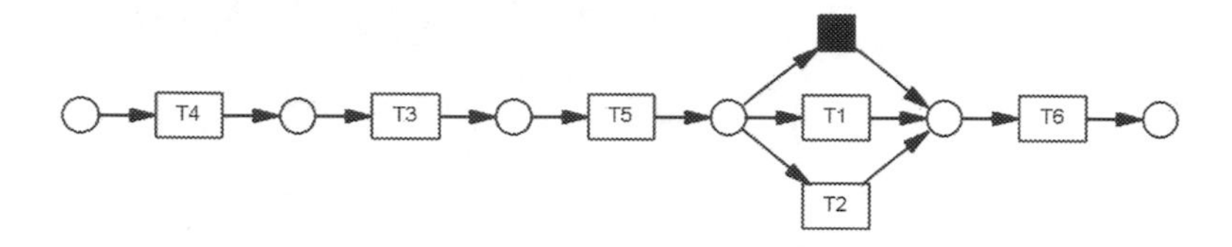

Fig. 5. A process model with an invisible task

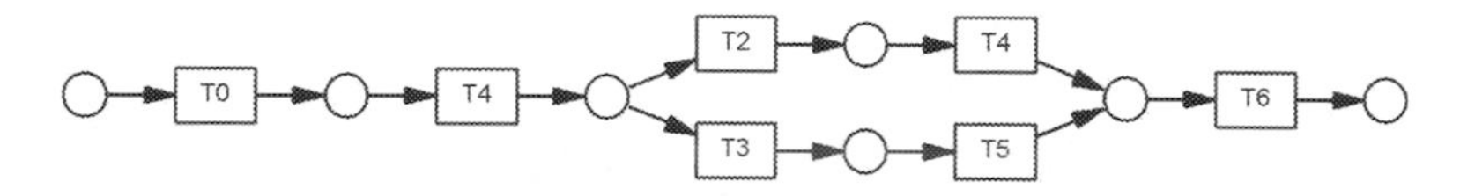

Fig. 6. A process model with duplicate tasks

non-free choice. Task T3 and Task T4 share a same place (i.e., the place *P1*), while they both have another input place which is not an input of the other. Clearly such construct is difficult to mine since the choice is non-local and the mining algorithm has to 'remember' all earlier events.

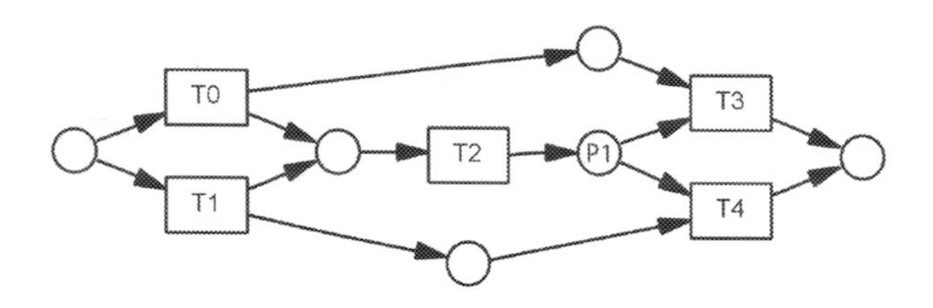

Fig. 7. A process model with a non-free choice

4.4 Arbitrary Loop

Arbitrary loop means a loop that has more than one exits or more than one entrances. An example of arbitrary loop is the process model in Figure 8. In this model, there is a loop (task T1, task T3 and task T5). There are two entrances (task T0 and task T7) to this loop. Besides, there are two exits from this loop (task T4 and task T6). It is difficult to mine the arbitrary loop because the loops can be entered or exited from more than one ways.

4.5 Short Loop

In a process, it may be possible to execute the same task arbitrary times. If this happens, this typically refers to a short loop in the corresponding model. Figure 9 is an example of a process model with short loop. In this model, task T4 is in a

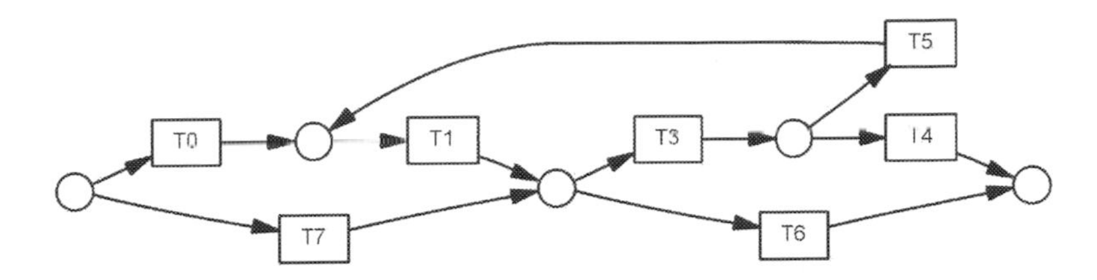

Fig. 8. A process model with an arbitrary loop

short loop. That is, task T4 can be executed arbitrary times. For more complex processes, mining loops is far from trivial since there are multiple occurrences of the same task in a given trace.

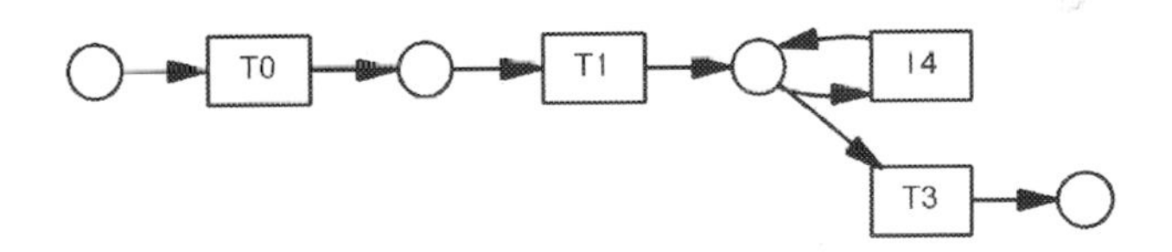

Fig. 9. A process model with a short loop

4.6 Nested Loop

A petri net with nested loop is such petri net where there exists a pair of loops, and these two loops share part of the tasks. For instance, a process model with a nested loop is in Figure 10. There are two loops in this process model: one is constituted with task T1, task T2 and task T6; the other is constituted with task T2 and T7. Clearly it is not easy to mine the nested loop constructs, since these loops can jump arbitrarily from one to the other.

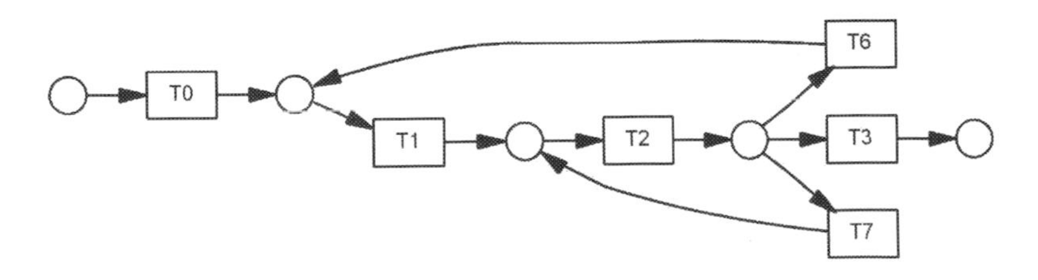

Fig. 10. A process model with a nested loop

5 Evaluation

In this section, we report the experimental evaluation. We managed to conduct two experiments: one is designed for evaluating the selected features in Section 3; the other is designed for evaluating the proposed universal significant reference model set.

The experiments are performed on BeehiveZ[9] using a PC with Intel(R) Core(TM) i7-2600@3.40GHz CPU, 8GB RAM.

5.1 The Datasets

Table 1. Dataset Properties

Dataset	Size	Average			Minimum			Maximum		
		#transitions	#places	#arcs	#transitions	#places	#arcs	#transitions	#places	#arcs
Artificial	270	6.100	6.244	13.233	2	3	4	13	14	30
Boiler	108	7.222	7.639	14.694	3	4	6	12	11	24
Trains	243	16.024	14.679	32.629	6	6	12	36	32	72

To construct a comprehensive benchmark for evaluating process mining algorithms, both artificial and real-life models are included in our empirical study. The basic properties of these models are summarized in Table 1. The following paragraphs briefly introduce the sources of the datasets and summarize the benchmark results.

The Artificial Dataset. Artificial process models (total 270 models) are collected from other papers, the SAP reference models, and also manually created by authors.

The Boiler Manufacturer Dataset. We have collected 108 real-life process models from Dongfang Boiler Group Co., Ltd. Dongfang Boiler is a major supplier of thermal power equipment, nuclear power equipment, power plant auxiliaries, chemical vessels, environmental protection equipment and coal gasification equipment in China (with more than 33% market share).

The High Speed Railway Dataset. We have collected 243 real-life process models from Tangshan Railway Vehicle Co., Ltd. (TRC). TRC is the oldest enterprise in China's railway transportation equipments manufacturing industry, with its predecessor constructing China's first railway line from Tangshan to Xugezhuang in 1881 during the westernization movement at the end of the Qing Dynasty. Today the company has built the train that has achieved the highest speed of Chinese railways - 394.3km/h.

5.2 Evaluation on Feature Selection

In the experiment on evaluating the feature selection, we compared our selected 6 features with the 48 features proposed in [1]. We select a fraction of process models (10%, 20%, 30%, 40%, and 50% repectively) from each datasets as training set to obtain the regression model, then recommend the most suitable process mining algorithm for the remaining models by applying the regression analysis.

Accuracy. To evaluate the accuracy, we first empirically evaluate the test models, and for each model, determine the algorithm that produces the mined model with maximum similarity (we called it the actual best algorithm). Then we apply our regression model to each of these test models without performing any empirical evaluation nor calculating the similarities. For each test model, we also determine the algorithm that has the highest estimated similarity (from the regression model), and it is called the estimated best algorithm. If the actual and estimated best algorithms are identical, we consider this as correct.

Figure 11(a), 11(b) and 11(c) show the accuracy result of Artificial, Boiler and Trains, respectively. With the size of training models increasing, the accuracy of recommendation is generally increasing. Overall, regression analysis with our selected 6 features can produce more accurate recommendation result than the one with 48 features. Besides, since the Trains dataset is more complex than the other two datasets, most process mining algorithms do not perform well on this dataset. This bad performance makes these process mining algorithms perform similar on this dataset (all perform not so well). Thus, a process model in Trains dataset can have several best algorithms. This leads to a better recommendation accuracy on the Trains than the other two datasets (as a process model could have several best algorithms, the framework could have a better chance to recommend a suitable algorithm). With the recommendation accuracy higher in Trains dataset than the other datasets, accuracy gap between the recommendation with 6 features and the one with 48 features in Trains dataset lower than the other dataset, as shown in Figure 11(c).

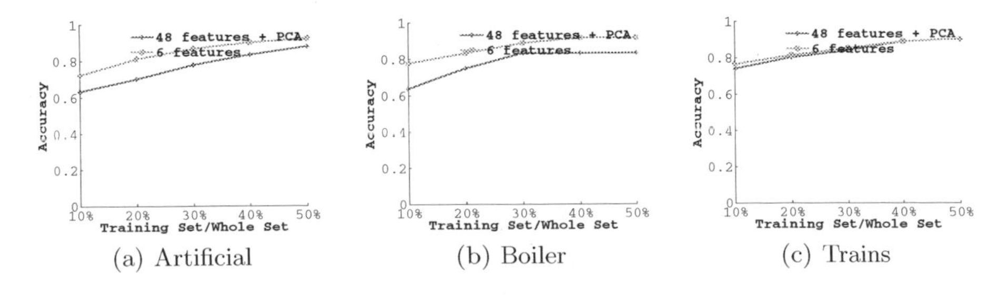

(a) Artificial (b) Boiler (c) Trains

Fig. 11. Accuracy Result on Evaluating Feature Selection

Efficiency. In the experiment, we contrast the feature extraction time between the 48 features and the 6 features. Figure 12(a), 12(b) and 12(c) show the speed result of Artificial, Boiler and Trains, respectively. Generally speaking, for the same model set and the same proportion training set, time cost of extracting 6 features is less than extracting 48 features. Besides, as the process models in Trains dataset are relatively complex, the time cost in Trains data set are more than other datasets. Meanwhile, the time cost of the Boiler dataset is least among these three datasets, because number of process models in Boiler is least.

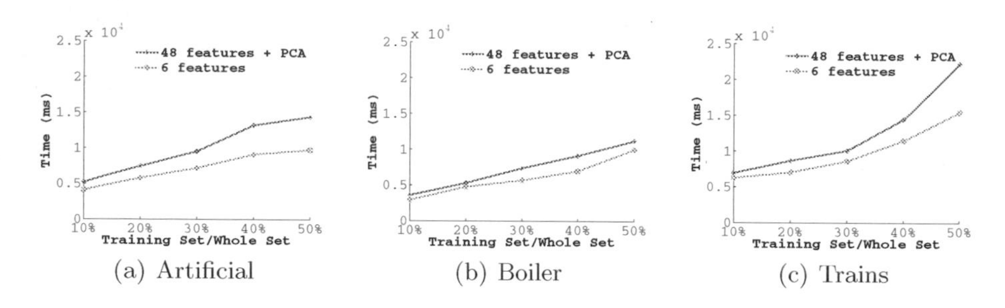

(a) Artificial (b) Boiler (c) Trains

Fig. 12. Time Result on Evaluating Feature Selection

5.3 Evaluation on Universal Significant Reference Model Set

In the experiment on evaluating universal significant reference model set, we compare our proposed universal significant reference model set (we call it *URM* for short) with the reference models selected from the each datasets (We call it *ORM* for short). In the *ORM*, we select one third process models of each datasets as the reference models. And, since the reference models in *ORM* are different regarding each dataset, we have to train its corresponding reference models for each dataset. In the *URM*, since the reference models are independent with the datasets, the models in the universal significant reference model set can be trained only once and be used without extra training for each different dataset.

Efficiency. The experiments result on speed is shown in Table 2.

For *ORM*, the time cost is composed of 3 parts: 1. *mining* (i.e. the empirical evaluation on one third of the models), 2. *training* (i.e. deriving the regression model), and 3. *recommending* (i,e. applying the regression model to recommend the best suitable algorithm).

For *URM*, the time cost is only composed of recommending. This is because that the models in *URM* only needs to be mined and trained once globally. And when recommending, the training result of *URM* can be just used without paying extra time on mining and training once again. Therefore, compared with the *ORM*, the time cost of *URM* is much smaller as shown in Table 2.

Table 2. Time Cost on Evaluating the Universal Significant Reference Model Set

Dataset	ORM(s)				URM(s)
	Mining	Training	Recommending	All	Recommending
Artificial	2789	83	14	2886	10
Boiler	1393	31	11	1435	10
Trans	18722	42	22	18786	15

Accuracy. The accurate criteria in this experiment is the same with the experiment on evaluating the feature selection: namely, for each test model, to determine whether the estimated best algorithm is the same with the actual best algorithm.

Table 3 shows the accuracy result of the *ORM* and *URM*. Overall, the results show that the recommendation using *URM* are more accurate than the recommendation using *ORM*.

Table 3. Accuracy on Evaluating the Universal Significant Reference Model Set

Dataset	Size	ORM		URM	
		#Correct	Accuracy	#Correct	Accuracy
Artificial	180	158	87.78%	166	92.22%
Boiler	72	60	83.33%	66	91.67%
Trains	162	146	90.12%	154	95.06%

6 Conclusion and Future Work

While some previous effort has been made to efficiently select suitable process mining algorithms by training from reference models, the performance is highly dependent on the quality of reference process models.

In this paper, we propose a universal significant reference model set. Furthermore, we carefully study the model features defined in [5], and propose 6 features that can capture the model rediscoverability more effectively and efficiently.

From the experiments, we can see our proposed universal significant reference model set has a reliable performance on recommending suitable process mining algorithms (all of the accuracy are above 90%), and greatly reduces the recommendation time. Besides, our proposed 6 process model features outperform the 48 features in characterizing model rediscoverability.

Our ongoing work includes further analyzing the models in this universal significant reference model set and the features that characterize the model's rediscoverabilities. And we hope to design a new process mining algorithm with better performance on model rediscoverability.

Acknowledgements. The work is supported by the National High-Tech Development Program (863 Plan) of China (No. 2012AA040904), the NSF of China (No. 61003099), and the Ministry of Education & China Mobile Research Foundation (MCM20123011).

References

1. Wang, J., Wong, R.K., Ding, J., Guo, Q., Wen, L.: Efficient Selection of Process Mining Algorithms. IEEE Transactions on Services Computing 6(4), 484–496 (2013), doi:10.1109/TSC.2012.20
2. Wang, J., Jin, T., Wong, R.K., Wen, L.: Querying business process model repositories - A survey of current approaches and issues. World Wide Web 17(3), 427–454 (2014)
3. Wang, J., Tan, S., Wen, L., Wong, R.K., Guo, Q.: An empirical evaluation of process mining algorithms based on structural and behavioral similarities. In: Proceedings of the 27th Annual ACM Symposium on Applied Computing, pp. 211–213. ACM (2012)
4. Wang, J., Wong, R.K., Ding, J., Guo, Q., Wen, L.: On recommendation of process mining algorithms. In: IEEE 19th International Conference on Web Services (ICWS), pp. 311–318. IEEE (2012)
5. Mendling, J.: Metrics for process models. LNBIP, vol. 6. Springer, Heidelberg (2008)
6. de Medeiros, A.A., Gnther, C.W., Weijters, A.J.M.M., van der Aalst, W.M.: Towards an evaluation framework for process mining algorithms. Beta, Research School for Operations Management and Logistics (2007)
7. Van der Aalst, W., Weijters, T., Maruster, L.: Workflow mining: Discovering process models from event logs. IEEE Transactions on Knowledge and Data Engineering 16(9), 1128–1142 (2004)
8. Van der Aalst, W.M., Weijters, A.J.M.M.: Process mining: A research agenda. Computers in Industry 53(3), 231–244 (2004)
9. Jin, T., Wang, J., Wu, N., La Rosa, M., ter Hofstede, A.H.M.: Efficient and accurate retrieval of business process models through indexing. In: Meersman, R., Dillon, T.S., Herrero, P. (eds.) OTM 2010. LNCS, vol. 6426, pp. 402–409. Springer, Heidelberg (2010)
10. Wang, J., He, T., Wen, L., Wu, N., ter Hofstede, A.H.M., Su, J.: A behavioral similarity measure between labeled Petri nets based on principal transition sequences. In: Meersman, R., Dillon, T.S., Herrero, P. (eds.) OTM 2010. LNCS, vol. 6426, pp. 394–401. Springer, Heidelberg (2010)
11. Wen, L., van der Aalst, W.M., Wang, J., Sun, J.: Mining process models with non-free-choice constructs. Data Mining and Knowledge Discovery 15(2), 145–180 (2007)
12. Wen, L., Wang, J., van der Aalst, W.M., Huang, B., Sun, J.: Mining process models with prime invisible tasks. Data & Knowledge Engineering 69(10), 999–1021 (2010)
13. de Medeiros, A.K.A., Weijters, A.J., van der Aalst, W.M.: Genetic process mining: An experimental evaluation. Data Mining and Knowledge Discovery 14(2), 245–304 (2007)
14. Weijters, A.J.M.M., van der Aalst, W.M., De Medeiros, A.A.: Process mining with the heuristics miner-algorithm. Technische Universiteit Eindhoven, Tech. Rep. WP, 166 (2006)
15. Van der Aalst, W.M., Rubin, V., van Dongen, B.F., Kindler, E., Gnther, C.W.: Process mining: A two-step approach using transition systems and regions. BPM Center Report BPM-06-30, BPMcenter. org. (2006)
16. Dijkman, R., Dumas, M., García-Bañuelos, L.: Graph matching algorithms for business process model similarity search. In: Dayal, U., Eder, J., Koehler, J., Reijers, H.A. (eds.) BPM 2009. LNCS, vol. 5701, pp. 48–63. Springer, Heidelberg (2009)
17. Zha, H., Wang, J., Wen, L., Wang, C., Sun, J.: A workflow net similarity measure based on transition adjacency relations. Computers in Industry 61(5), 463–471 (2010)

A Systematic Methodology for Outpatient Process Analysis Based on Process Mining

Minsu Cho[1], Minseok Song[1,*], and Sooyoung Yoo[2]

[1] Ulsan National Institute of Science and Technology, 50 UNIST-gil, Eonyang-eup, Ulju-gun, Ulsan, Republic of Korea, 689-798
{mcho,msong}@unist.ac.kr

[2] Seoul National University Bundang Hospital, 82 Gumi-ro 173 beon-gil, Bundang-gu, Seongnam-si, Gyeonggi-do, Republic of Korea, 463-707
yoosoo0@snubh.org

Abstract. A healthcare environment has been important due to the increase in demand for medical services. There have been several research works to improve clinical processes such as decreasing the waiting time for consultation, optimizing reservation systems, etc. In this paper, we suggest a method to analyze outpatient processes based on process mining. Process mining aims at extracting knowledgeable information from event logs recorded in information systems. The proposed methodology includes data integration, data exploration, data analysis, and discussion steps. In the data analysis, process discovery, delta analysis, and what-if analysis using performance analysis results are conducted. To validate our method, we conduct a case study with a tertiary general university hospital in Korea.

Keywords: Process Mining, Healthcare, Case Study, Business Process.

1 Introduction

A healthcare environment has been important due to the increase in demand for medical services caused by population aging and improved standards of living. For this reason, not only the high-quality consultation from medical progress but the optimal clinical process from the business process perspective has been provided to patients [2].

In the past, all information related to patients was written by hand on chart. In contrast, at present hospitals have several information systems used to record patients' information. In general, PMS (Practice Management System), EMR (Electronic Medical Record), CPOE (Computerized Physician Order Entry), and PACS (Picture Archiving Communication System) are used in hospitals. By using these information systems, all processes are stored from visiting the hospital to going home and those are used as information to improve the clinical processes. In addition, hospitals can manage the patients and provide the better service at a lower cost by analyzing the data [7].

* Corresponding author.

C. Ouyang and J.-Y. Jung (Eds.): AP-BPM 2014, LNBIP 181, pp. 31–42, 2014.

Research on the outpatient process has been performed by several researchers. Lindley and Jansson tried to decrease the waiting time for consultation using a queuing model [5, 6]. Fries and Marathe suggested the optimal reservation method by testing lots of reservation systems [4]. Wang studied in decreasing both idle time and consultation time [12]. However, these papers didn't propose a systematic analysis method or analyze based on the data. In addition, those were not process level analyses. Therefore, we suggest a method to analyze the outpatient process based on process mining.

Process mining aims at extracting knowledgeable information that is related to processes from the event log recorded in an information system [10, 11]. Process mining consists of three categories, which are discovery, conformance, and extension. Discovery produces a process automatically from the event log. Conformance represents to find the discrepancies between the log and the model. Lastly, extension is to improve the process model by using other attributes. Process mining can be applied to various industry fields such as manufacturing, medical IT devices, and call-centers. Process mining is also used in the healthcare environment. However, it is difficult to analyze the data because there are unstructured processes (spaghetti processes) in healthcare. In this paper, we suggest a systematic methodology from data integration and exploration to data analysis and discussion. In addition, we have a case study with a tertiary general university hospital, which is a fully digitized hospital in South Korea, to validate our methodology.

The paper is structured as follows. Section 2 describes the works related to outpatient processes and process mining while section 3 introduces the overall methodology for analyzing outpatient processes. In section 4, we will validate our methodology with a case study. Lastly, our conclusion and future works will be described in section 5.

2 Related Works

2.1 Process Mining

The purpose of process mining is to extract process-related information by analyzing event logs that are stored in information systems. The event logs for analysis usually have been produced in PAISs (Process Awareness Information Systems) such as Workflow management, ERP, and so on.

Process mining consists of three types: Discovery, Conformance, and Enhancement. Normally, process mining focuses on the discovery process model from the event logs. Conformance checking compares the process model with the observed behavior in the event logs. Lastly, enhancement is to improve the process model using other attributes such as time, and cost.

Additionally, process mining has an advantage that is applicable in various fields. A lot of process mining techniques have been applied in many business processes such as healthcare, ECM, port, and manufacturing.

2.2 Researches in Healthcare Outpatient Process

Researches for improving outpatient processes have been conducted for the last several years. Lindley did a research to reduce the patient's waiting time by using a queuing model [6]. Additionally, Fries and Marathe compared reservation systems for consultation to find the optimized number of patients in each slot [4]. Robinson and Chen conducted a research to find the best reservation system considering a given order and the number of patients [8]. In this way, there have been lots of researches to improve the outpatient processes through optimized reservation methods or waiting time reduction methods.

Process mining researches for outpatient processes also have been conducted in various ways. Mans et al. suggested a methodology for analyzing healthcare processes by applying process mining, which was based on hospital data from the Netherlands [7]. Moreover, they addressed a way of extracting a social network model and doing performance analysis. Alvaro and Diogo conducted a pattern analysis of outpatient processes to extract the standard behavior and the unusual behavior [1]. In addition, there was a research work to analyze compliance rules of healthcare processes for skins cancer patients [3].

Even though there are many researches to analyze outpatient processes, the methodology for analyzing outpatient care processes with process mining is not structured and systematic yet. However, it is crucial to develop a scientific methodology to analyze outpatient data easily and readily. For this reason, we propose an analysis framework for outpatient data.

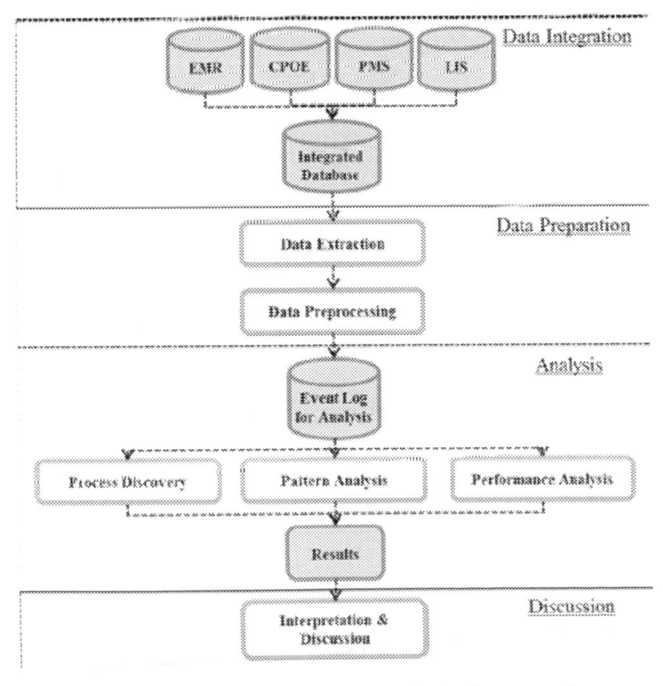

Fig. 1. Outpatient process analysis framework

3 A Methodology for Applying Process Mining in Outpatient Process Analysis

In this section, we describe a methodology to analyze outpatient processes by applying process mining. Figure 1 represents the overall outpatient process analysis framework. It consists of 4 parts: data integration, data preparation, analysis, and discussion.

3.1 Data Integration and Data Preparation

In order to store the records of patients, resources and so on, various information systems are utilized at hospitals. The HIS (Hospital Information System) consists of several applications, such as PMS, EMR, CPOE, PACS, LIS (Laboratory Information System). Therefore, an integrated database needs to be built by integrating records from each information system. The data integration is progressed by considering the EMR data as basic and adding information recorded in CPOE and LIS. In this part, a patient-oriented integrated database should be made based on the records in each information system. Since the same patient ID is shared among each database, it is easy to make an integrated database. Figure 2 shows the relationship between information systems.

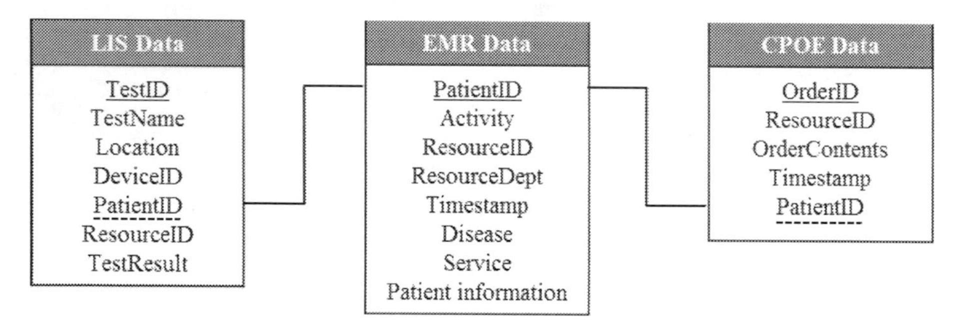

Fig. 2. Data structures from several information systems

After the data integration, the required data are extracted from the integrated database. In order to extract data, an analyst should think about what he or she wants to know or analyze. For example, this analysis is intended for outpatients' data and needs to follow the form of event logs to apply the analysis techniques of process mining. Thus, not only key information such as patient ID, activity, resource ID, and timestamp, but also additional information such as the type of patient is extracted.

3.2 Data Analysis

A. Process Discovery and Delta Analysis

Control-flow discovery aims at extracting the process automatically from the event log. It is to produce a process from the event sequences of each case. There are several mining techniques such as alpha-algorithm, heuristic mining, fuzzy mining, and so on. Alpha-algorithm is a basic mining method using causal dependencies between all activities in the event log. In addition, petri-net is a representative model of alpha-algorithm and it consists of three tuples: places, transitions, arcs. The alpha-algorithm has some advantages that can be applied to several performance analyses. However, it includes the outliers in the process and can't represent certain parts of patterns such as loops. To overcome these weaknesses, various mining techniques have emerged such as heuristic mining and fuzzy mining.

In the hospital, there is a standard outpatient process from domain experts. It is called the outpatient care mega-process. However, the actual logs can't be matched well with the standard one, although it is an optimal process for the patients. Accordingly, we need to do a delta analysis that compares the care mega-process with the process from the event log. Delta analysis is used to understand the difference between the process from the log and the care mega-process [9]. Visualization techniques are frequently used to highlight the difference. That is, coloring on the process model shows the difference points between two processes.

Unlike delta analysis, the difference between two processes needs to be quantified for direct comparison. This is because it can be easily understood how they are different. Therefore, we suggest a method to calculate the matching rate between the meta-care process and the event log.

Definition 1 (Matching rate). Let L be an event log over A and M_p be a Meta-care process that has same activities with L. $A = \{a_1, a_2, a_3, a_4, \cdots\}$.$|a_1 > a_2|$ is the number of times a_1 is directly followed by a_2 in L:

$$|a_1 > a_2| = \sum_{\sigma \in L} L(\sigma) \times |\{1 \leq i < |\sigma| \mid \sigma(i) = a_1 \wedge \sigma(i+1) = a_2\}|$$

$|a_1 \sqsupset a_2|$ is the number of times a_1 is directly followed by a_2 in L when the directly followed relationship from a_1 to a_2 is also in M_p:

$$|a_1 \sqsupset a_2| = \begin{cases} |a_1 > a_2| & \mathit{if}\ a_1 > a_2\ \mathit{exists\ in}\ M_p \\ 0 & \mathit{else} \end{cases}$$

Matching rate is how many directly follows in L are matched with directly follows in M_p:

$$\text{Matching rate (\%)} = \frac{\sum_{1 \leq j < |A|} \sum_{1 \leq k < |A|} |a_j \sqsupset a_k|}{\sum_{1 \leq j < |A|} \sum_{1 \leq k < |A|} |a_j > a_k|} \times 100$$

By using the above formulas, the matching rate between an event log and an outpatient care mega-process can be calculated. It represents the number of the matched directly-followed relationships divided by the number of all directly-followed relationships. From the matching rate, we can easily check the difference between two processes. For example, Table 1 represents the directly-followed relationship from an artificial event log. In the table, 'From' means precedent tasks and 'To' means following tasks. The value in each cell indicates the frequency from a task to another task. If a care process consists of a sequence of activities A, B and C, the cells from A to B and from B to C are on the flow. Thus, the matching rate becomes 50 %(=(5+4)/(2+5+1+0+2+4+0+1+3)*100).

Table 1. An example of directly-followed relationship matrix

Frequency		To		
		A	B	C
From	A	2	**5**	1
	B	0	2	**4**
	C	0	1	3

B. Pattern Analysis

As we introduced earlier, the extracted outpatient process has a form of spaghetti in the healthcare environment [10]. Unlike a lasagna process, a spaghetti process is unstructured, so we can't understand the patterns easily. Accordingly, pattern analysis is needed to figure out the overall flow from visiting the hospital to going out. To do the analysis, Performance Sequence Diagram Analysis in ProM is used.

By applying the technique, we can get the frequent patterns in the outpatient process. However, not only the pattern analysis from the event log that includes all data, but also the analysis depending on the attributes of patients, is needed to give information according to the type of patients. We suggest attributes that can divide the patients. Firstly, patients can be divided based on whether they have been admitted to the hospital previously or not. In general, the patterns of patients who visit the hospital for the first time are long and complex because they should have several tests, register their personal information and so on. On the contrary, the patterns of patients who are return-visitors to hospital are relatively simple. Therefore, this attribute should be considered. In addition, other criteria are shown in Table 2. By using these attributes, the pattern analysis can be extended in various ways.

Table 2. Attributes of patients which can be used for pattern analysis

Attributes	Description
New or Returning	Whether they have been visited to the hospital previously or not
Advanced reservation or not	Whether they have an advanced reservation or not
Medical insurance or not	Whether they have a medical insurance or not

C. Performance Analysis and Simulation

The performance analysis using process mining can be applied to many subjects in hospital process analysis. It has a wide range of application areas, such as the information about how long a patient stayed at the hospital, the time for each activity, the frequency of each resource, work time, and so on. Available performance analyses are shown in Table 3.

Also, a what-if analysis can be done based on the result of these kinds of performance analysis. The existing simulation analysis was based on the information provided by hospital staff, but it is possible to build an actual simulation model by using the result of data analysis.

Table 3. Subjects of Performance Analysis

Subjects of analysis	Description
The overall time for out-clinic	The time consumed from visiting hospital to going out
The time for each activity	The time consumed between the previous activity and the activity
The frequency of each activity per head	How many times the activities are performed for one person
The work time for each activity by resources	The time consumed between the previous activity and the activity of each resource
The frequency for each activity by resources	How many times the resources perform the activities
The hourly frequency of patients	How many patients are in the hospital hourly

3.3 Discussion

Lastly, it is necessary to have a discussion with experts in the hospital. Based on the comparison results between the event log and the standard process, the outpatient care mega-process may be modified. This is because some parts that were not recorded in the event log may be meaningless, even though those are included in the standard process. By enhancing the process, we can derive the new outpatient care mega-process. In addition, the route information applied the pattern analysis results can be provided to the patients who visit a hospital for the first time because they might not be able to know the proper process. Furthermore, simulation results may be applied such as adjusting the number of resources and changing the reservation system for consultation since it is difficult to try directly in the real world.

4 Case Study

In this section, we validate our outpatient process analysis methodology that was suggested in section 3. For the analysis, an event log for outpatients was extracted from the HIS at SNUBH (Seoul National University Bundang Hospital). The event log contained 15 tasks: consultation, test, payment, etc. It included information about resources, departments, patient types, etc. The log contained the patients visiting times in May of 2012. The summary information was as follows:

- About 120,000 cases (patients that were treated)
- About 700,000 events (activities performed for these patients)
- 15 different tasks (e.g. consultation, test, payment, etc.)

4.1 Process Discovery and Evaluation of Outpatient Process

To evaluate the standard process model in the hospital, we derived process models from the log using several control-flow discovery techniques and compared them with the standard one. We applied the heuristic mining, the fuzzy mining, and the comp mining which we developed in the project. The comp mining is a method that uses only the number of directly-followed relationship among activities in an event log.

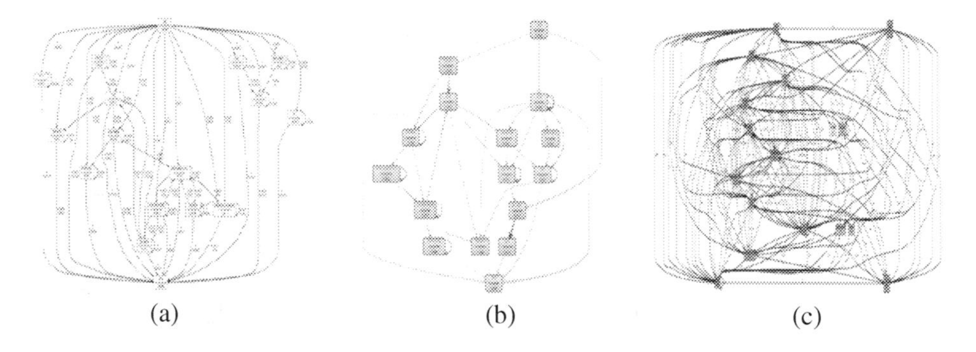

Fig. 3. The models using (a) Heuristic Mining, (b) Fuzzy Mining and (c) Comp Mining

Figure 3 shows the derived models that show actual process flow in the hospital. The heuristic mining result and the fuzzy mining result showed major flows in the outpatient process, while the comp mining result showed all possible paths among the activities including less-frequent flows. The most frequent flow is from *consultation registration* and *consultation,* which occurred about 64,000 times. The flows happened more than 10,000 times are as follows.

- *Consultation registration → Consultation*
- *Test → Consultation registration*
- *Consultation → Payment*
- *Consultation scheduling → Payment*
- *Payment → Payment*
- *Test → Test registration*
- *Payment → Test registration*
- *Test registration → Test*

- *Payment → Outside-hospital prescription printing*
- *Payment → Treatment*

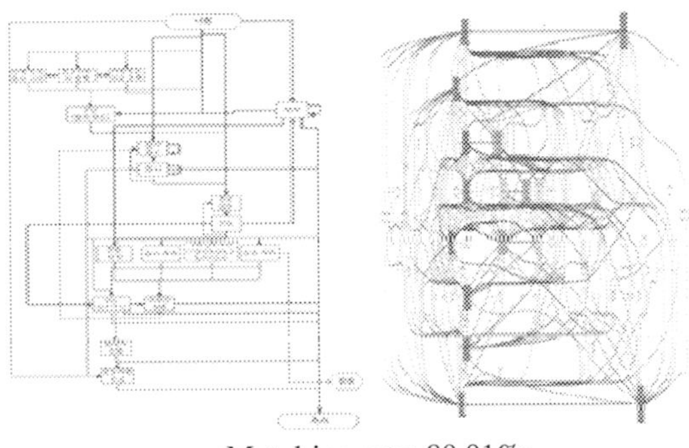

<Matching rate: 89.01%>

Fig. 4. Comparison with the care mega-process model in the hospital

Figure 4 (a) is the standard model in the hospital and Figure 4 (b) shows the differences between the standard model and the comp mining result. In the figure, the red lines and the green ones represent non-matching and matching flows respectively. We calculated the matching rate between two models which is 89.01%. The comp mining result shows that the actual outpatient care processes were very complex compared to the standard model, since the standard one contains only the important flows among all possible movements. However, during the discussion with medical professionals we were not able to find any undesired patterns in the derived model, i.e. the process is well-managed by the hospital.

4.2 Process Pattern Analysis

In the hospital, there are several types of patients. Among them, there are patients who visit the hospital for the first time ('new patients') and patients who are return-visitors to hospital ('returning patients'). To analyze the differences between the types of patients, we derived process models and performed the process pattern analysis considering patient types.

Figure 5 shows the derived process models. The 'returning patients' enrolled a consultation room and got a consultation as soon as they visited the hospital, while the 'new patients' were registered an outside referral document for the first task. The pattern analysis shows that 'new patients' stayed longer than 'returning patients' in the hospital. The most frequent patterns for each patient type are as follows:

- 'New patients': RORD→CR→C→P
- 'Returning patients': CR→C→RC→P→OPP

(RORD: Registration of an outside referral document,
CR: Consultation registration, C: Consultation, P: Payment,
RC: Reservation for consultation, OPP: Outside-hospital prescription printing)

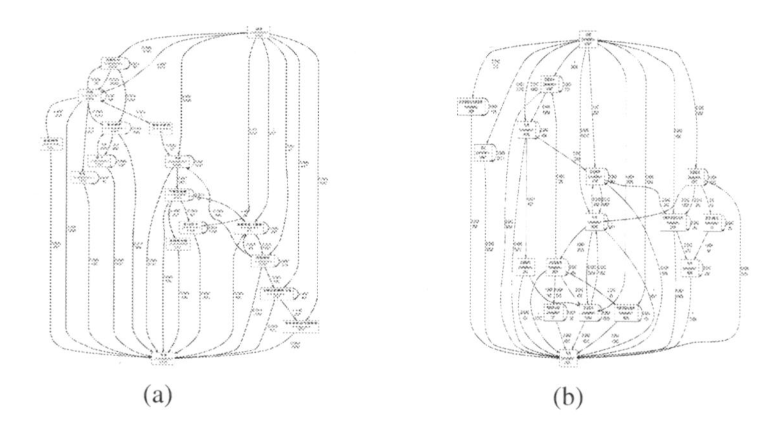

Fig. 5. The models of (a) 'New Patients' and (b) 'Returning Patients'

The pattern analysis result was used to build the smart healthcare system in the hospital. Patients can use a smartphone to find their route that is recommended based on the pattern analysis result.

4.3 Performance Analysis and Simulation

We performed basic performance analysis from the log and used the result to make business process simulation model. To make a simulation model, several parameters were used such as a process model, an arrival rate of a case, and time information of activities, etc. For the simulation, we calculated the execution time for the activities by applying the performance analysis, which is one of process mining techniques.

Based on the performance analysis result, we made a simulation model for a doctor. Using the simulation model, we analyzed the change of consultation waiting time according to the increase of patients. Figure 6 shows how the waiting time became longer according to the increase. In the figure, the 10% increase of patients makes the biggest increase in waiting time which is about 25%. Thus, we suggested less than 10% of increase if they need the change of the number of patients.

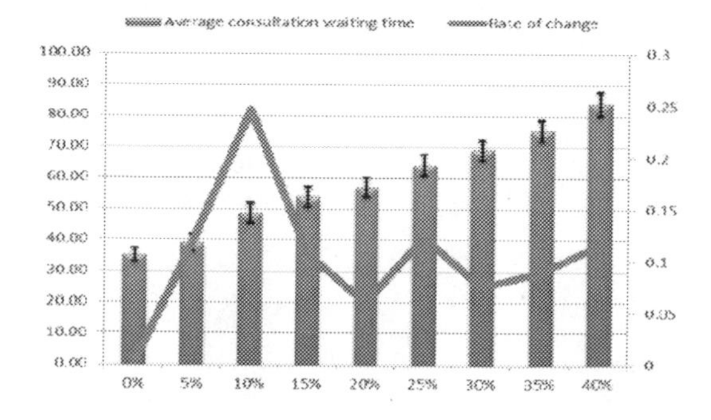

Fig. 6. The result for consultation waiting time

5 Conclusion

In this paper, we suggested a methodology for analyzing outpatient processes based on process mining. In addition, we conducted a case study based on the outpatients' event log in SNUBH. We derived the process model and compared it with the standard model in the hospital. In addition, we analyzed the process patterns according to patient types and conducted performance analysis and make a simulation model using the analysis results.

According to the result of comparing the event log and their standard process model, the matching rate was as 89.01%. That is, they relatively well understood workflows of outpatients and the process was well-managed by the hospital. Using the performance analysis result, we generated the simulation model. The simulation shows that the 10% increase of patients makes the largest change in consultation waiting time. Thus, we recommended less than 10% of increase. In addition, we extracted the process models and analyzed the process patterns according to patient types. The most frequent pattern of each patient type was discovered. (e.g. 'Returning patients': CR→C→RC→P→OPP) The patterns were used to build a smart guidance app in the smart healthcare system in the hospital. Patients can use a smartphone to find their route that is recommended based on the pattern analysis result.

For the future, we will conduct a research for the segmented outpatient process analysis. In general, each doctor has specialties in certain diseases, so they have their own standard process for better services. Accordingly, a low-level process analysis will be performed to make more generalized standard process.

Acknowledgments. This work was supported by the Industrial Strategic Technology Development Program funded by the Ministry of Trade, Industry & Energy (MOTIE, Korea) (No. 10040142).

References

1. Álvaro, R., Diogo, R.F.: Business process analysis in healthcare environments: A methodology based on process mining. Information Systems 37(2), 99–116 (2012)
2. Anyanwu, K., Sheth, A., Cardoso, J., Miller, J., Kochut, K.: Healthcare enterprise process development and integration. J. Res. Pract. Inf. Technol. 35(2), 83–98 (2003)
3. Binder, M., et al.: On Analyzing Process Compliance in Skin Cancer Treatment: An Experience Report from the Evidence-Based Medical Compliance Cluster (EBMC^2). In: Ralyté, J., Franch, X., Brinkkemper, S., Wrycza, S. (eds.) CAiSE 2012. LNCS, vol. 7328, pp. 398–413. Springer, Heidelberg (2012)
4. Fries, B., Marathe, V.: Determination of Optimal Variable-Sized Multiple-Block Appointment Systems. Operations Research 29(2), 324–345 (1981)
5. Jansson, B.: Choosing a Good Appointment System: A Study of Queues of the Type (D/M/1). Operations Research 14, 292–312 (1966)
6. Lindley, D.V.: The Theory of Queues with a Single Server. Proceedings Cambridge Philosophy Society 48, 277–289 (1952)

7. Mans, R., Schonenberg, H., Leonardi, G., Panzarasa, S., Cavallini, A., Quaglini, S., van der Aalst, W.M.P.: Process Mining techniques: an application to stroke care. Stud. Health Technol. Inform. 136, 573–578 (2008)
8. Robinson, L.W., Chen, R.: Scheduling doctors' appointments: Optimal and empirically-based heuristic policies. IIE Trans. 35(3), 295–307 (2003)
9. van der Aalst, W.M.P.: Business Alignment: Using Process Mining as a Tool for Delta Analysis and Conformance Testing. Requirements Engineering 10(3), 198–211 (2005)
10. van der Aalst, W.M.P.: Process Mining: Discovery, conformance and enhancement of business processes. Springer, Heidelberg (2011)
11. van der Aalst, W.M.P., Schonenberg, M.H., Song, M.: Time Prediction based on Process Mining. Information Systems 36(2), 450–475 (2011)
12. Wang, P.P.: Static and Dynamic Scheduling of Customer Arrivals to a Single-Server System. Naval Research Logistics 40, 345–360 (1993)

Measuring Patient Flow Variations: A Cross-Organisational Process Mining Approach

Suriadi Suriadi[1], Ronny S. Mans[2], Moe T. Wynn[1], Andrew Partington[3], and Jonathan Karnon[3]

[1] Queensland University of Technology, Australia
{s.suriadi,m.wynn}@qut.edu.au
[2] Eindhoven University of Technology, The Netherlands
r.s.mans@tue.nl
[3] University of Adelaide, Australia
{andrew.partington,jonathan.karnon}@adelaide.edu.au

Abstract. Variations that exist in the treatment of patients (with similar symptoms) across different hospitals do substantially impact the quality and costs of healthcare. Consequently, it is important to understand the similarities and differences between the practices across different hospitals. This paper presents a case study on the application of process mining techniques to measure and quantify the differences in the treatment of patients presenting with chest pain symptoms across four South Australian hospitals. Our case study focuses on cross-organisational benchmarking of processes and their performance. Techniques such as clustering, process discovery, performance analysis, and scientific workflows were applied to facilitate such comparative analyses. Lessons learned in overcoming unique challenges in cross-organisational process mining, such as ensuring population comparability, data granularity comparability, and experimental repeatability are also presented.

Keywords: Process mining, data quality, patient flow, data mining.

1 Introduction

The impact of an aging population and the surge in lifestyle diseases on health service demand globally is well documented [1,2]. Consequently, policy makers are under pressure to rationalise and justify expenditure on public services. There is a plethora of literature that provides best practices guidelines on the provision of cost-effective care; however, such recommendations are implemented in heterogeneous contexts, resulting in continued variances in costs and quality [3]. To achieve greater efficiencies in healthcare delivery, this paper attempts to measure and explain the variations in clinical practices across different hospitals.

The main research question, according to the clinicians, is to identify to what extent cross-hospitals variations exist and why they exist. In this case study, we propose the use of process mining techniques to identify the similarities and differences in the patient flows of four South Australian (SA) hospitals. Process Mining [4] combines data mining and process analysis techniques. These techniques

C. Ouyang and J.-Y. Jung (Eds.): AP-BPM 2014, LNBIP 181, pp. 43–58, 2014.

can be applied to the healthcare datasets to discover the pathways that patients traversed within hospitals. The specific case study objectives are to compare process models and logs between different hospitals, to measure performance differences between hospitals, and to identify sub-groups (i.e., cluster of cases) that can explain the variations in patient flows.

This paper provides example results from our case study that highlight key variations between the hospitals and offer insights from within the data as to why these variations exist. We also detail the challenges and lessons learned with a focus on data challenges specific to hospital data. This paper extends the preliminary findings from our earlier work [5] where a different set of process mining techniques were applied to a different dataset (which contains similar patients data) for comparisons between different hospitals. This paper, by contrast, attempts to quantify the similarities and differences between models and logs and to explain the variations using clustering and decision tree analysis. Furthermore, a detailed performance comparison is provided for all four hospitals against key milestones and national target guidelines. By comparing the activities undertaken and evaluating the performance differences, we gain insights into how process behaviours differ between hospitals (e.g., under-performing or over-achieving) which can be used to, in consultation with clinicians, elicit information about improvement opportunities.

2 Data Preparation

Data Source. The South Australian Health Department maintains a Clinical Reporting Repository (CRR) which stores and collates electronic medical record for each patient from a number of publicly-funded health providers. Each hospital also maintains its own information systems for managing operating theatres and tracking patient transfers between physical wards.

The data extracted from these systems captures activities related to Emergency Department (ED) care, inpatient care, pathology tests, and procedures performed. While the data was extracted from different systems, they are comparable: the schema for the CRR data is already standardised, while data from individual hospital is recorded based on industry standards, such as the Australian modified, International Classification of Diseases and Health Related problems (ICD-10-AM) [6] and other similar standards.

Scope. Our study excludes cases whereby (1) the patients involved were transferred-in from another hospital (where they had another admission immediately prior to the ACS related ED presentation), (ii) the patients had a prior ED presentation within the preceding 12 months, or (iii) the cases were not sufficiently documented (e.g., missing pathology tests). The above exclusions left us with an event log of 9997 cases, 263712 events, and 64 different event classes.

Conceptual Model. The UML class diagram of the extracted data is provided in Figure 1. An individual case (the `'case'` class) consists of a unique case identifier and the related patient identifier. A patient can be involved in multiple cases (e.g., repeated visits to an ED). A patient has many attributes, such as the patient identifier and the gender (see the `patient` class). Within a case, multiple events may occur (see the `'event'` class). The `'event'` classs stores the related activity name ('concept:name'), its type ('lifecycle:transition'), the timestamp ('time:timestamp'), and the corresponding resource ('org:resource').

An event can occur within an Emergency Department (ED) (see the `'ED'` class), such as the presentation of a patient at an ED (the `'ED presentation'` class) and the consultation of a patient with a doctor (the `'doctor seen'` class). In this paper, we refer to these events as `EDpres` and `DrSeen` respectively.

An event may refer to procedures (the `'procedure'` class) applied to, or tests undertaken for, patients. For tests, we distinguish between a blood test (the `'blood test'` class) and a medical image (the `'medical imaging'` class).

Finally, events may relate to an `'admission/discharge'`. This class consists of an admission request event (the `admission request` class) - henceforth referred to as `AdmReq`, a patient ED discharge event (the `'ED discharge'` class), a patient admission to, and discharge from, inpatient care event (the `'inpatient care admittance'` and `inpatient care discharge` class respectively), and the the movement of patient between wards (the `'ward admission'` event). We henceforth denote the `'inpatient care admittance'` event as `AdmAsInpat`, and the `'ward admission'` event as `Ward` event.

The `'ED discharge'` event has a corresponding departure status, including discharge to ward (`ED_Disch_Ward`), discharge to home (`ED_Disch_Home`), transfer to another hospital (`ED_Disch_TransOH`), and discharge to a ward within ED (`ED_Disch_AdmW/ED`). Note that additional attributes exist for some events.

Data Cleaning. To discover process knowledge, a correct temporal ordering of events within a log is important. Thus, the accuracy of timestamp is crucial. To check the accuracy of timestamp, we performed a few checks, such as the execution of the "Timestamp Issue Detector" plugin of ProM 6 [7] to detect duplicate timestamps, timestamps with different accuracy, and timestamps that may be outliers within a trace. Also, we visually inspected the log for other timestamp issues. Table 1 summarises the identified timestamp issues.

Table 1. Identified timestamp issues in the event log

Issue	% of total events
Timestamp accurate to a day, e.g., 01/01/2012 00:00:00	0.1%
Timestamp accurate to an hour, e.g., 01/01/2012 15:00:00	10.1%
Timestamp accurate to a minute, e.g., 01/01/2012 15:34:00	37.2%
Same timestamp for 'start' and 'complete' event	84.1%
Start and complete events with timestamp difference < 60 secs	84.8%
Groups of different events with the same timestamp	88.1%

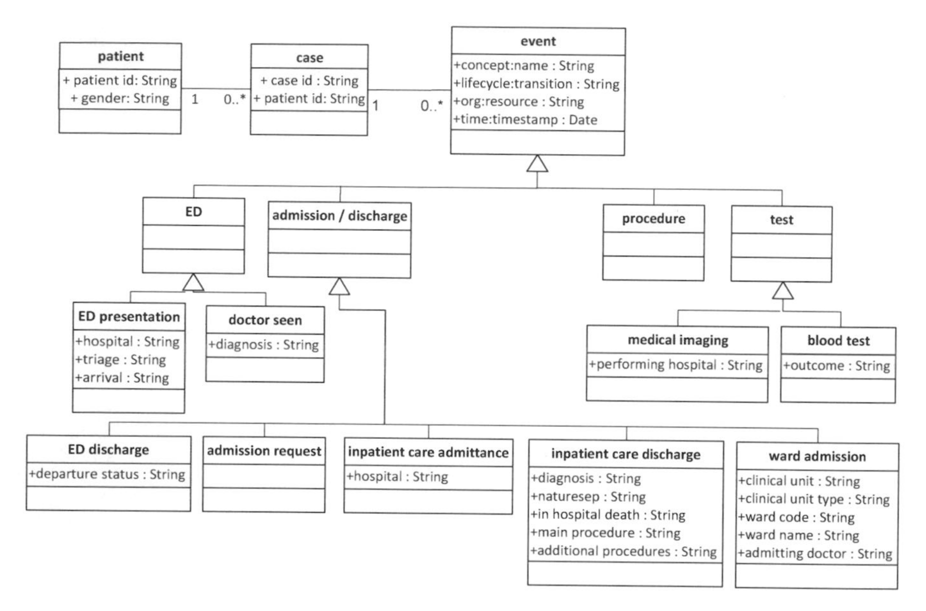

Fig. 1. Conceptual model of the event log used in this case study

The issues mentioned in Table 1 calls for data pre-processing to discard inaccurate timestamp information. We applied three filters to the data: (1) we filtered out cases that contain at least one event whose timestamp is not accurate in terms of a second, (2) with the exception of the `Ward` event, we only keep the 'complete' events because the rest of the activities in the log have the same "start" and "complete" timestamps, (3) we only use data for patients with *chest pain* as their diagnosis (i.e., R07 as the ICD-10-AM diagnosis code). The resulting *filtered* log contains 3040 cases, 69872 events, and 33 event classes.

We split the log into three cohorts based on their ED discharge status (see Table 2): Cohort 1 (C1) contains cases with patients admitted to a hospital ward `ED_Disch_Ward` or `ED_Disch_AdmW/ED`), Cohort 2 (C2) contains cases with patients discharged home (i.e., `ED_Disch_Home`), and Cohort 3 (C3) contains cases with patients transfered to another hospital (i.e., `ED_Disch_TransOH`). There are 3002 cases captured by these three cohorts (the remaining 38 cases contained other ED discharge status not considered in this paper).

As shown in Section 3, certain analyses require a more complete population. Thus, we prepared a second log derived by applying the last filter only (i.e., only cases with R07 diagnosis code, without the other timestamp filters). We call this log the *unfiltered* log, it contains 6037 cases, 142830 events, and 64 event classes.

3 Measuring Variations across Hospitals: Results

In this case study, data and process analysis techniques were used to measure and quantify differences between clinical practices across hospitals. To enable

Table 2. The distribution of cases per hospital for each cohort (*filtered* log)

Cohort	Hospital 1	Hospital 2	Hospital 3	Hospital 4
C1 (Admitted)	667	719	240	392
C2 (Home)	205	264	140	231
C3 (Transferred)	137	2	1	4

proper comparative analysis, we filtered the dataset to obtain similar cohorts of patients across the four hospitals. Section 3.1 presents our findings from the process discovery task where process models are discovered for each hospital. The similarity amongst the models is then measured by replaying a log from one hospital against the model from another hospital. Section 3.2 then compares the performance of each hospital based on key patient flows milestones (e.g., presentation and ED discharge) and against the national guidelines. This allows us to explore the potential links between the overall patient flows behaviours and their respective performance measures (e.g., throughput time, length of stay). The analysis results confirm that significant variations in process behaviour and process performance exist across these four hospitals. To better understand the reasons behind these variances, we carried out clustering analysis to explore alternative ways of clustering patient cohorts based on patient characteristics given in the dataset as well as process characteristics derived from the dataset. Section 3.3 presents the preliminary findings from this clustering analysis.

3.1 Process Comparison

For comparison, process models capturing patient flows need to faithfully reflect the behaviours captured in the log, i.e., the fitness of the models need to be high. The following approach was used to obtain the model for each hospital.

Patient flows include activities executed both at the ED and the ward (in-hospital). Accordingly, we split the log into two sublogs: one captures the patient flows in the ED (i.e., all events from `EDpres` up to one of the events representing ED discharge activity), while the other captures the patient flows in the ward (i.e., all events after the ED discharge point). This approach may simplify process discovery as fewer events are seen in each sublog.

Then, for each sublog, we discovered the respective process models by using the Heuristics Miner [8], the Passage Miner [9], and the Fuzzy miner [10] available in the ProM Tool. To avoid clutter in the discovered models, we considered only those events with a relative occurrence of at least 5%. Based on the discovered models, a consolidated Petri net model was created by hand. Through the application of conformance checking plugin [11], we then assessed the cost-based fitness value of the created Petri net model. The above-described activity was repeated until the manually-created Petri net model achieves a fitness value of above 0.8 (sufficient to state that the model captures the majority of the behaviours seen in the log).

Finally, the ED and in-hospitals models obtained were merged. Due to space limitation, we only show the the ED part of the process model in Fig. 2. The red oval-circles in this figure highlight the differences in the control flow between these four hospitals (identified manually) as described below.

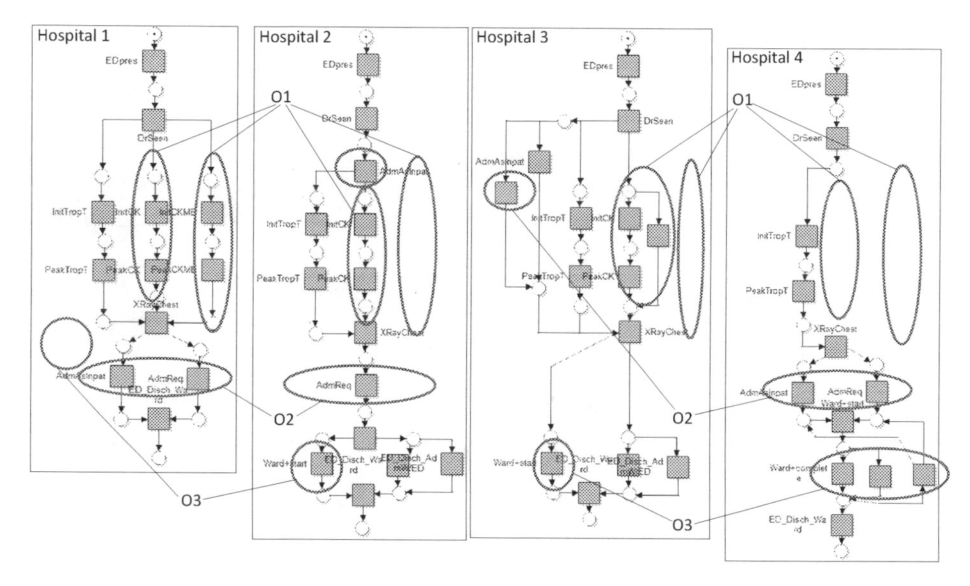

Fig. 2. Discovered ED patient flows - differences coded as 'O1', 'O2', and 'O3'

Observation 1 (O1) : Blood tests are performed by all hospitals; however, there is a variation in terms of the tests performed. In all hospitals, *Troponin T (TropT)* tests (`InitTropT` and `PeakTropT`) are performed. For the *Creatine Kinase (CK)* tests (`InitCK` and `PeakCK` events), hospitals 1 (H1) and hospital 2 (H2) performed it regularly; hospital 3 (H3) performed this test in an optional manner, and hospital 4 (H4) did not perform this test at all. The *Creatine Kinase - MB (CKMB)* test (`InitCKMB` and `PeakCKMB` events) is only performed by hospital 1.

Observation 2 (O2) : Within H1 and H4, the `AdmReq` event and the `AdmAsInpat` event are executed after the X-ray chest examination. For H2 and H3, this also holds true, but the `AdmAsInpat` event already took place after the `DrSeen` event. Nevertheless, for H3, the `AdmReq` event did not take place.

Observation 3 (O3) : Within H2 and H3, the `Ward` (start) event seems to already take place before the patient has left the ED (i.e., `ED_Disch_Ward` or `ED_Disch_AdmW/ED`). Within H4, it is even possible that the patient has already been admitted to several wards (i.e., both start and complete `Ward` events) before the any of the two ED discharge events took place.

To 'quantify' the extent of the observed differences, we replayed each hospital's log on the Petri net models of four hospitals. The cost-based fitness metric [11] for each replay was calculated, the results of which are shown in Table 3 (due to space limitation, we only show results for C1 patients).

Note that the timestamps of blood test events (e.g., InitTrop, TroughTrop, and PeakTropT) are typically similar. So, their ordering in the log may not necessarily follow the same ordering as they happened in reality, i.e., they follow a partial order. Put it differently, one may also say that the exact ordering of these events are *not important* as they typically happened more or less at the same time. Thus, in order to remove artificial ordering of these events and to not unfairly punish traces that did not meet the exact ordering of these events in the discovered model, we created a standardised ordering of these events such that they all occurred according to the following order: `Init`, `Trough`, and `Peak`. This ordering is chosen as it has been discovered that for most cases, the events follow this order. For example, if we see the following order of events `PeakCK`, `PeakTropT`, `InitTropT`, and `InitCK`, these are reordered into `InitTropT`, `PeakTropT`, `InitCK`, and `PeakCK`. In this way, the fitness of the models increases and they can be better compared amongst hospitals.

As shown in Table 3, the model for H1 has the lowest fitness compared with the logs for the other hospitals. For the models of H2, H3, and H4, their fitness compared with the logs of the other hospitals are not that distinct. We thus conclude that the patient flows at H1 seem to be distinct from others. The cause for this can be twofold: (1) within H1, the patient is not already admitted to a ward before being discharged from the ED, i.e., the `Ward` event in the logs of H2, H3, and H4 does not exist in the model of H1, (2) within H1, all three blood test (*TropT*, *CK*, and *CKMB*) are performed whereas this is not the case for other hospitals, i.e some transitions in the model of H1 cannot be mapped to any of the events in the logs of the other hospitals.

Finally, the models of H2 and H3, and their associated logs, are quite comparable, informing us that the patient flows of H2 and H3 are comparable. The only differences are that the `AdmReq` event of H2 is not present in the model of H3, and that the `AdmAsInpat`, `InitCK`, and `PeakCK` events are optional in H3.

Table 3. Fitness values - replay of each hospital's log on each hospital's model

	Process Model (Petri Net)			
Hospital log	H1	H2	H3	H4
Hospital 1 (H1)	0.918	0.756	0.745	0.749
Hospital 2 (H2)	0.651	0.861	0.836	0.748
Hospital 3 (H3)	0.586	0.784	0.847	0.726
Hospital 4 (H4)	0.611	0.725	0.77	0.871

3.2 Performance Analysis

We have so far measured the similarity in the patient flows across different hospitals. Such an exercise is necessary to understand how the process may impact on performances. To this end, we conduct a performance analysis for each of these hospitals.

For the patient flows in cohort C1, some important milestones exist. First, a patient presents at an ED (EDpres), then the patient is seen by a doctor (DrSeen). An admission request can then be made (AdmReq) before the patient is discharged from the ED (either ED_Disch_Ward) or ED_Disch_AdmW/ED). Finally, the patient is then discharged home from the ward (iDisch_Home). Table 4 summarises the average time, and the corresponding standard deviation, spent between these milestones.

We also performed the same analysis on *unfiltered* log. This analysis was performed to check whether the presence of events with a different timestamp accuracy has a negative impact on the performance analysis. Our results show that there are no large differences in the results between the obtained metrics for the *filtered* and *unfiltered* logs. This could be due to the relatively 'balanced' distribution of events with accurate and inaccurate timestamps in our data set which somehow lessens the impact of timestamp accuracy on aggregated results. However, had the population been skewed towards events with inaccurate timestamp, the differences in the results could have been more pronounced. Regardless, for this case study, we can conclude that the performance results on *filtered* log (Table 4) are still valid for the population in the *unfiltered* log.

In addition to the above milestones, the operation of ED within Australian hospitals needs to meet certain national targets. These targets dictate the minimum percentage of cases that should not exceed (1) the maximum time for a patient to be discharged from ED counted from the presentation time, and (2) the maximum time elapsed for a patient to consult a doctor based on their triage category. Because we need to capture the percentage of cases that met (or did not meet) the national standards, we performed our analysis on *unfiltered* log - the results of which are shown in Table 5.

Table 4. Milestones analysis results

Milestones Analysis (*filtered* log)	H1 avg (std.dev)	H2 avg (std.dev)	H3 avg (std.dev)	H4 avg (std.dev)
EDpres to DrSeen	0.30 (0.45)	0.25 (0.48)	0.24 (0.33)	0.19 (0.24)
DrSeen to AdmReq	4.14 (2.6)	4.1 (3.68)	3.91 (4.85)	2.35 (1.47)
AdmReq to ED_Disch_Ward	8.9 (7.99)	7.51 (7.57)	11.38 (7.57)	4.23 (2.82)
AdmReq to ED_Disch_AdmW/ED	not seen	8.42 (4.75)	4.59 (3.29)	not seen
ED_Disch_Ward to iDisch_Home	37.68 (58.08)	32.40 (52.32)	18.30 (33.12)	52.80 (65.52)

Explaining Performance Differences. A few notable differences in performance are observed, e.g., for the AdmReq to ED_Disch_Ward milestone, and for

Table 5. National ED target analysis results

Measure (*unfiltered* log)	Target	H1	H2	H3	H4
ED stay < 4 hours	90%	17%	22%	22%	24%
Triage 1 (Immediate)	100%	100%	100%	not seen	100%
Triage 2 (<10 mins)	80%	64%	81%	83%	75%
Triage 3 (<30 mins)	75%	77%	58%	76%	67%
Triage 4 (<60 mins)	70%	71%	50%	60%	82%
Triage 5 (<120 mins)	70%	not seen	50%	100%	100%

the `ED_Disch_Ward` to `iDisch_Home` milestone, the difference between H3 and H4 is quite high. These differences can be explained by looking into the waiting times and the working times for these milestones. H3 has a larger average time in between the `AdmReq` and `ED_Disch_Ward` milestone because its `ED_Disch_Ward` event has a substantially longer waiting time than for H4 (i.e., 10.41 hours vs. 1.51 hours respectively). Furthermore, H4 has a larger average time between the `ED_Disch_Ward` and `iDisch_Home` milestone than H3 because H4 had large waiting times for the `iDisch_Home` and the `Ward` (start) events (i.e., 17.16 hours for H4 and 4.54 hours respectively). This is in contrast with H3 with average waiting time of 4.54 hours and 2.32 hours respectively.

For national ED target analysis, we can see that across all hospitals, only about a quarter of all ED cases met the 4-hour ED length-of-stay rule, with H1 performing the worst. Furthermore, for the critical Triage category 2 patients, only H2 and H3 conform to the national standard (i.e., 80% of all Triage 2 cases to be consulted by doctors within 10 minutes). While further analysis is needed to establish any correlation, recall that our earlier process comparison analysis (Section 3.1) concludes that H1's process seems to be distinct from others, while highlighting the the fact the patient flows of H2 and H3 to be quite similar. Such an analysis is useful as it guides our analysis to see what H2 and H3 do differently then H1 and H4 which allow them to meet the national rules.

3.3 Clustering Analysis

As our intention is to compare patient flows across four hospitals, we decided to group cases based on their ED discharge status (e.g., discharged to home or admitted to hospital) and then compare each cohort across four hospitals (cf. Section 3.1 and Section 3.2). However, such a top-down approach to data splitting may have ignored alternative natural groupings of cases that may explain the reasons behind the variations. For instance, are there similarities in patient characteristics that are well-known by the clinicians that could explain these differences? Alternatively, are there any process characteristics (e.g., the number of times a certain activity is undertaken, the number of events in a case) that could explain the differences? To this end, we employ a clustering technique to

(1) obtain clusters of cases when one considers both the flow perspective and clinically-relevant attributes, and (2) measure the similarity in the population between our earlier exit-point-based grouping and the clusters' population.

Clustering is performed on a case log: in addition to available case attributes (e.g., the triage category of each case, the hospital in which a case occurred, and clinical unit type), the log is enriched with derivable process-related attributes, i.e., (1) the key exit point of patients from ED and the hospitals (e.g., `ED_Disch_Home`), (2) the case duration, (3), total number of events in a case, (4) the relative timestamp of these ED departure events (i.e., the time from the start of a case to the occurrence of an ED departure event). The k-means clustering algorithm available from the WEKA tool [12] was applied to obtain four clusters. Four clusters were chosen as we would like to examine whether the four main ED exit points could be a good approximate measure for clustering cases.

Once we obtained the clusters, we measure the intersection of the population from each cluster with the population of each of the cohorts we used in our analysis. For clarity, we split the C1 population into two cohorts: C1.1 (for `ED_Disch_AdmW/ED` patients), and C1.2 (for `ED_Disch_Ward` patients). If our assumption about the use of ED exit point as cases separation point is correct, then we expect the population of each of our four cohorts (C1.1, C1.2, C2, and C3) to strongly match the population of exactly one of the four clusters, and vice versa. The results of our analysis are shown in Fig. 3.

The graph on the right in Fig. 3 captures the proportion of a cohort's population that are captured by the four clusters discovered. The population of the 'Home' cohort strongly matches the population of cluster 3 (over 90% of cases from the C2 intersect cluster 3). Similarly, the population of the C3 cohort also strongly matches the population of cluster 0 (92% of C3 exist in cluster 0). However, the population of the C1.2 cohort is split amongst multiple clusters (e.g., 44% of C1.2 exist in cluster 1, while the rest are split amongst cluster 2 and cluster 0). Similar observation applies to the 'Admit ED' (C1.1) cohort.

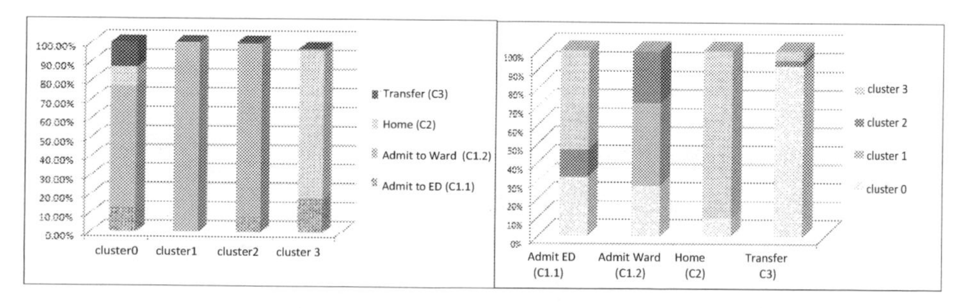

Fig. 3. The percentage of cluster population that exists within each cohort (left) and the percentage of cohort population that exist within each cluster (right)

Conversely, Fig. 3 (left) captures the proportion of a cluster's population that also exists within each of the four cohorts. From this figure, we see that cluster 3 contains mostly the C2 cohort (78%). Cluster 1 and cluster 2 are strongly linked with 'Admit Ward' (C1.2) cohort (100% and 92% respectively). However, the population of cluster 0 contains population of all four cohorts.

Key observations from these results: (1) the C2 ('Discharge to Home') cohort indeed forms one natural group as there is a strong one-to-one mapping between the clusters and the cohorts, (2) the C3 ('Transfer to Other Hospital') cohort do share similar characteristics among not only themselves (the majority of them falls within one cluster), but also with population from other cohorts (the population of C3 are mostly in cluster 0, but cluster 0 also consists of population from C1.1, C1.2, and C2 cohorts), (3) the C1.2 ('Admit to Ward') and C1.1 ('Admit to ED') cohorts contains further diversity that need further separation, while at the same time, these two cohorts also intersect similar clusters, notably cluster 0 and cluster 1, thereby suggesting that there are overlapping characteristics.

The above analysis suggests that the clusters that were formed were not strictly based on ED exit points, and that additional attributes need to be considered to further separate the population of our original cohorts. We also applied a decision tree analysis [13] to extract the key features that differentiate these clusters. The results showed that in addition to ED exit points, other differentiating features for cohorts include the number of events within an episode, hospital name, and triage category. We plan to explore the possible explanations behind these features in order to understand the variations among clusters as part of the future work.

4 Challenges and Lessons Learned

We now summarise challenges and lessons learned from this case study.

Data Granularity. The data collated from different organisations and extracted from different systems typically contain data attributes with varying degree of granularity. We have discussed in detail specific issues related to the granularity of timestamps in Section 2 whereby certain events were recorded with very coarse granularity (e.g., day-level accuracy), while others were recorded up to second-level accuracy. The key issue faced is to understand the impact of such varying degree of granularity on our analysis, and how to address it.

Our case study showed that for certain analyses, e.g., process comparison, timestamp granularity needs to be most accurate. We find the use of *case-level* filtering to remove all cases in which at least one of the events do not meet the second-level accuracy to be crucial as the existence of even an event with a coarser timestamp granularity within a trace can easily distort the actual ordering of events. For other analyses, such as performance analysis, uniform timestamp granularity may not matter that much. The results of our experiment (see Section 3.2) demonstrate that, contrary to the common belief, the performance analysis results using the *filtered* and *unfiltered* log are very similar. Hence, a key lesson is that, for important attributes like event timestamps, we need to first assess the varying degree of granularity that could be found in the dataset, then evaluate the possible impact on the different types of analyses to be conducted and filter the data appropriately to suit the analysis being conducted.

Ensuring Population Comparability. Patient flows data is complex: it captures not only the highly-varied sequences of clinical activities which make up patient flows, but also a wide range of clinically-relevant information (e.g., types of tests and procedures applied to patients and the types of wards in which patients are being treated). Given such complexity, it can be difficult to determine appropriate ways to split the data into natural groupings for comparison purposes. A key challenge here is how to group cases such that each group is mostly homogeneous, and thus, yield usable comparison results.

In our case study, we used two approaches: we grouped the cases based on the hospital name and ED discharge status and we also used clustering techniques and decision tree analysis to create clusters based on clinically-relevant attributes and derived process characteristics. We then compared the cluster results with the original cohorts. A key lesson learned here is that clustering techniques can be used to discover inherent clusters of cases informed purely by the features in the dataset while decision tree analysis can be used to extract the key discriminating features of each cluster. Furthermore, the discriminating features found from the decision tree analysis may assist in explaining variations that exist within cohorts made by using only hospital name and ED discharge status.

Simplifying Spaghetti Models. The complexity of logs obtained from various organizations is likely to result in highly-complex spaghetti process models, if one were to approach process discovery naively. Through this case study, we reflect on a number of techniques that can be used to address this situation.

- *Splitting Discovery of Sub-processes.* By splitting the process discovery approach based on natural sub-processes within a process, we can remove the complexity in the discovered model by disentangling certain paths. In our case study, various blood tests occur both during the ED and the inpatient sub-processes. If we were to discover just one process model for both sub-processes, each blood test event will be captured as one activity, resulting in many cross-arcs as arcs will be drawn from activities that exist within the ED and inpatient sub-processes. If we were to treat these sub-processes separately, the same activity will then be 'disentangle' into two distinct activities (e.g., in a Petri Net model, the same activity will be captured as two transitions), thus, simplifying the model as arcs are now more 'contained': blood test activities conducted during the ED and in-patient sub-processes only result in arcs drawn within their respective region. By subsequently combining these two sub-process models, a cleaner model can be obtained.
- *'Sequentializing' Events whose Strict Ordering are Unimportant.* As described in Section 3.1, we observed that the ordering of certain events do seem to happen almost at the same time, resulting in a conclusion that the exact ordering of these events to be inconsequential. A process model generated without further data pre-processing will result in an artificial ordering of these events which may either be incorrect (if they are ordered sequentially), or result in a complex model (if they are arranged in a parallel manner, resulting in many arcs). To simplify the discovered model, and to avoid unfairly punishing the fitness of a model due to its artificial events ordering, we repaired

the log by 'sequentializing' such events. Consequently, the resulting process model also shows them as a series of sequential activities. This results in fewer arcs in the model, and replaying of these events on such a model will also fit (avoiding the punishment of a model's fitness unfairly).
- *Manual Model Creation.* Despite the two techniques described above, the use of existing process model discovery algorithms (such as Passage Miner and Heuristic Miner) will still likely to result in a complex model due to the inherent limitations of each algorithm. We took the approach of observing the models discovered from each algorithm, and merging these models (manually) such that the most complete behaviours of the process, as seen in the log, can be captured within a single model. In this case study, by manually created our own process model, informed by the ones discovered through existing algorithms, we managed to obtain a model with a much improved fitness value.

Experiment Repeatability. Cross-organizational process mining often requires one to perform the same analysis repeatedly using a log from a different organisation in each iteration. Manually conducting such repetition is tedious and time consuming. In this case study, we used Rapid Miner[1], a scientific workflow tool, that has been extended with a series of process mining algorithms, to conduct many of those repetitive experiments. Our experience with the scientific workflow tool is very positive as it improved the speed in which analysis results can be obtained: the replaying of four logs from different hospitals over the four models created for each hospital, over the three main cohorts (i.e., a total of 48 replays) can be achieved in an almost fully-automated manner (manual efforts are required to describe the workflow and to set the parameters). Most importantly, however, is that by saving the scientific workflow used for a certain experiment, we can always *repeat* the same experiment over and over again. This is a very useful capability, especially when we need to work with logs that have been changed slightly from experiment to experiment.

5 Related Work

Given the benefits of process mining, researchers are increasingly applying it in the domain of healthcare. Webster [14] introduced general ideas of process mining and the potential/benefits from its application to healthcare. In our earlier work [5], we reported findings from a literature survey which highlighted the research gaps in terms of the lack of comparative analysis across multiple hospitals. Mans et al. [15] identified twelve studies related to the application of process mining in different health care processes, e.g., the gynecological oncology process in a Dutch hospital [16], the emergency process in a public hospital in Portugal [17], the process of an inpatient's journey from admission to discharge in an Australian public hospital [18], and the process of breast cancer treatment in a hospital in Belgium [19].

[1] www.rapidminer.com

Many of the existing case studies focus on the discovery of processes using datasets from one hospital [18,20–22]. An exception is this study [16] where data from 368 patients diagnosed with 'first-ever ischemic stroke' from four Italian hospitals were analysed to discover the treatment procedures of the patients. Process discovery and conformance analysis techniques were used to analyse skin cancer treatment process in [21]. Data challenges encountered in healthcare datasets were also described in [15, 17, 23]. The use of clustering techniques as part of the process mining analysis for healthcare was discussed in [24]. This paper reports on a comparative case study using a much larger cohort of patients which enables us to study the effect of high process behaviour variations.

6 Conclusion

This paper presented a process mining case study which analysed data sets from four hospitals to better understand the variations in clinical practices across these hospitals with a particular focus on process behaviours and their respective performance. Detailed comparative analysis was carried out using techniques such as process discovery, performance analysis, and clustering analysis. The results provide detailed insights into the key differences in process behaviours and the level of similarity across different hospitals. The performance analysis results also indicate the differences in performance across hospitals and highlight the need to understand the underlying factors for process delays. We presented key lessons learned from this case study, especially in terms of overcoming the challenges commonly encountered in cross-organisational process mining to ensure comparability of results. The insights gained from this case study are instrumental in guiding our subsequent analyses to explain reasons for the variations in patient flows, and most importantly, understand their impacts on clinical outcomes. As part of the future work, we need to incorporate existing process similarities quantification techniques [25, 26] to measure process variations across hospitals. The use of such techniques is needed when we need to compare a large number of process models and/or complex models.

Acknowledgement. This research is supported by the HCF Health and Medical Research Foundation grant for the project entitled *Reducing variation in clinical practice: a twin track approach to support improved performance* (2012-2014).

References

1. Lowthian, J.A., Curtis, A.J., Jolley, D.J., Stoelwinder, J.U., McNeil, J.J., Cameron, P.A.: Demand at the emergency department front door: 10-year trends in presentations. The Medical Journal of Australia 196(2), 128–132 (2012)
2. Odden, M., Coxson, P., Moran, A., Lightwood, J., Goldman, L., Bibbins-Domingo, K.: The impact of the aging population on coronary heart disease in the United States. The American Journal of Medicine 124(9), 827–833 (2011)

3. Runciman, W.B., Hunt, T.D., Hannaford, N.A., Hibbert, P.D., Westbrook, J.I., Coiera, E.W., Day, R.O., Hindmarsh, D.M., McGlynn, E.A., Braithwaite, J.: CareTrack: Assessing the appropriateness of health care delivery in Australia. The Medical Journal of Australia 197(2), 100–105 (2012)
4. van der Aalst, W., et al.: Process mining manifesto. In: Daniel, F., Barkaoui, K., Dustdar, S. (eds.) BPM Workshops 2011, Part I. LNBIP, vol. 99, pp. 169–194. Springer, Heidelberg (2012)
5. Partington, A., Wynn, M.T., Suriadi, S., Ouyang, C., Karnon, J.: Process mining for clinical processes: A comparative analysis of four Australian hospitals. Technical Report 66728, Queensland University of Technology (2013), http://eprints.qut.edu.au/66728 (last accessed March 4, 2014)
6. World Health Organization: International statistical classification of disease and related health problems - Tenth Revision (ICD-10), Geneva (1992)
7. Eck, M.: Timestamps Within Healthcare Process Mining Logs. Master's thesis, Eindhoven University of Technology, Eindhoven (2013)
8. Weijters, A.J.M.M., van der Aalst, W.M.P., Medeiros, A.K.A.: Process mining with the heuristic miner-algorithm. Technical report, Eindhoven University of Technology (2006)
9. van der Aalst, W.M.P.: Decomposing process mining problems using passages. In: Haddad, S., Pomello, L. (eds.) PETRI NETS 2012. LNCS, vol. 7347, pp. 72–91. Springer, Heidelberg (2012)
10. Günther, C.W., van der Aalst, W.M.P.: Fuzzy mining- adaptive process simplification based on multi-perspective metrics. In: Alonso, G., Dadam, P., Rosemann, M. (eds.) BPM 2007. LNCS, vol. 4714, pp. 328–343. Springer, Heidelberg (2007)
11. Adriansyah, A., van Dongen, B., van der Aalst, W.M.P.: Conformance Checking using Cost-Based Fitness Analysis. In: EDOC, pp. 55–64. IEEE (2011)
12. Witten, I., Frank, E., Hall, M.A.: Data Mining: Practical Machine Learning Tools and Techniques. Morgan Kaufmann (2011)
13. Quinlan, J.R.: C4.5: programs for machine learning. Morgan Kaufmann Publishers Inc., San Francisco (1993)
14. Webster, C.: EHR business process management: From process mining to process improvement to process usability. In: HSPI, Las Vegas, USA (2012)
15. Mans, R.S., van der Aalst, W.M.P., Vanwersch, R.J.B., Moleman, A.J.: Process Mining in Healthcare: Data challenges when answering frequently posed questions. In: Lenz, R., Miksch, S., Peleg, M., Reichert, M., Riaño, D., ten Teije, A. (eds.) ProHealth 2012 and KR4HC 2012. LNCS, vol. 7738, pp. 140–153. Springer, Heidelberg (2013)
16. Mans, R., Schonenberg, H., Leonardi, G., Panzarasa, S., Cavallini, A., Quaglini, S., van der Aalst, W.M.P.: Process mining techniques: An application to stroke care. In: MIE. Stud. in Health Tech. and Inf., vol. 136, pp. 573–578. IOS Press (2008)
17. Rebuge, A., Ferreira, D.: Business process analysis in healthcare environments: A methodology based on process mining. Inf. Syst. 37(2), 99–116 (2012)
18. Perimal-Lewis, L., Qin, S., Thompson, C., Hakendorf, P.: Gaining Insight from Patient Journey Data using a Process-Oriented Analysis Approach. In: HIKM 2012. CRPIT, vol. 129, pp. 59–66. ACS (2012)
19. Poelmans, J., Dedene, G., Verheyden, G., Van der Mussele, H., Viaene, S., Peters, E.: Combining Business Process and Data Discovery Techniques for Analyzing and Improving Integrated Care Pathways. In: Perner, P. (ed.) ICDM 2010. LNCS, vol. 6171, pp. 505–517. Springer, Heidelberg (2010)
20. Blum, T., Padoy, N., Feußner, H., Navab, N.: Workflow mining for visualisation and analysis of surgeries. J. of Comp. Assist. Rad. and Surgery 3(5), 379–386 (2008)

21. Binder, M., et al.: On Analyzing Process Compliance in Skin Cancer Treatment: An Experience Report from the Evidence-Based Medical Compliance Cluster (EBMC^2). In: Ralyté, J., Franch, X., Brinkkemper, S., Wrycza, S. (eds.) CAiSE 2012. LNCS, vol. 7328, pp. 398–413. Springer, Heidelberg (2012)
22. McGregor, C., Catley, C., James, A.: A Process Mining Driven Framework for Clinical Guideline Improvement in Critical Care. In: LEMEDS, pp. 35–46 (2011)
23. Lang, M., Bürkle, T., Laumann, S., Prokosch, H.: Process Mining for Clinical Workflows: Challenges and Current Limitations. In: MIE. Stud. in Health Tech. and Inf., vol. 136, pp. 229–234. IOS Press (2008)
24. Bose, R.P.J.C., van der Aalst, W.M.P.: Analysis of patient treatment procedures. In: Daniel, F., Barkaoui, K., Dustdar, S. (eds.) BPM Workshops 2011, Part I. LNBIP, vol. 99, pp. 165–166. Springer, Heidelberg (2012)
25. Mendling, J.: Metrics for Business Process Models - Empirical Foundations of Verification, Error Prediction, and Guidelines for Correctness. LNBIP, vol. 6. Springer (2008)
26. Dijkman, R., Dumas, M., van Dongen, B., Kaarik, R., Mendling, J.: Similarity of business process models: metrics and evaluation. Inf. Sys. 36(2), 498–516 (2011)

A Study on Geospatial Constrained Process Modeling Using UML Activity Diagrams

Guobin Zhu and Xinwei Zhu

International School of Software, Wuhan University, Wuhan, China, 430079
{gbzhu,xinwei.zhu}@whu.edu.cn

Abstract. This paper studies an extended process model under geospatial constraints using UML 2.0 activity diagrams. The geometry (*i.e.* point, line and polygon) models and geospatial relationship (*i.e.* measurement, sequential, and topology) models used in the paper are tailored from OGC and ISO/TC 211 standards with respect to requirements of process management. On the basis of these models, the paper analyzes how geospatial information restrains activities in process meta-model and how it acts on activities. Further, an UML profile is derived based on UML activity diagrams, and is validated by a case scenario. The contribution of this paper is marked as two points. First, the geometry models and geospatial relationship models proposed in this paper are based on abstract specification of international standards, making them better for understanding, sharing, and exchanging. Second, the geospatial constraints relied on geospatial relationship models adapt to most cases in process modeling, making them more generalizable and extendable.

Keywords: Geospatial Constraint, Process Modeling, UML, Activity Diagram.

1 Introduction

Over the past decades, Business Process Management (BPM) has become one of the abiding approaches to manage enterprises and create process-aware information systems (Van der Aalst, 2011). BPM has a broader scope: from process automation and process analysis to process management and the organization of work. On the one hand, BPM aims to improve operational business processes. For example, modeling and analyzing a business process with simulation leaves enterprise systems insensitive to external process variables. Although the model tries to describe an idealized version of reality, it is sometimes unable to adequately capture human behavior due to lack of descriptions of possible constraints against activity modeling.

In order to solve the issues mentioned above, many scholars shift to study using process context as a new paradigm of BPM in dealing with dynamic business process environments. In this paradigm, processes can be rapidly changed and adapted to a new external context (e.g., location, weather patterns, scales etc.). It is recognized that contextualizing processes provide more explicit consideration of the environmental setting of a process (Rosemann et al., 2008). Context in this research area is often focused to the term of locality (*e.g.* how to limit an activity to be triggered in a

C. Ouyang and J.-Y. Jung (Eds.): AP-BPM 2014, LNBIP 181, pp. 59–73, 2014.

specific location), and geospatial information (*e.g.* how do I choose a un-ruined road to get to a safe place when I am driving a car?).

Since geographic context can be treated as an illustration of generic contexts impacting business process and behaviors, it of course inherits importance of contexts described in (Rosemann et al., 2006, Rosemann et al., 2008). Meanwhile, as a special category of contexts, geographic contexts will hypothetically provide manifold benefits for business process management as follows:

- A geographical location embedded process model tends to be more objective for the sake of understandability.
- In the regular process modeling, geographical location seems to be treated as a constant variable in the process, resulting in that the influences of the location constraints are actually ignored. In the extended process modelling, the designers and the management can be aware of more deviations based on geographical context that takes place in the process. Also, it is easy to observe the work items and resources' behavior during the process execution, which helps to optimize the process based on the current status.
- The model with geographical information helps to identify the abstraction level to be a suitable one either too simple or too detailed, making a process flexible.

As existing literature on the influence of geographical location or information to the process model is rare, the related research discusses on this topic either in theory or by cases discussion. For example, (de Leoni et al., 2012) useed the map metaphor to visualize work items and resources in the context of the open-source workflow environment YAWL. It only mentioned that users could check geographical positions and distances based on a geographical map. However, its focus was not geographical information and therefore it did not discuss how the geographical information influences the process positively. In (Zhang et al., 2012), location constraints were expressed at the design time and help to tackle challenges during the runtime of the workflow schema. But location constraints presented in (Zhang et al., 2012)are lack of comprehensions.

However, the studies above are from the aspect of normal geographical location instead of geographical information. As the notion of latter one has a broader and semantic range than the former one. One pitfall of focus on the geographical location could be failed to support the complicated location-aware process model. For example, the 6 types of location relationship identified in (Decker et al., 2009) can not entirely cover all of the location relationships, therefore lack of comprehensions.

This paper summarizes possibilities of the geographical constraints for the process modeling. It analyzes how geographical information influences the process model, deriving an extension modeling method based on geographical information constraints. This tends to help designers and managers to concentrate on the objectives but not "desirable" behaviors. Thus, it will significantly address the behaviors of geographical location even geographical information in the normal modeling.

2 Background and Preliminary Knowledge

2.1 Workflow Process Meta-model

A meta-model is a model that defines a language for expressing a model, which is used to define the construct and rules of a semantic model (Van der Aalst, 2000). The workflow meta-model is used to describe the elements, their relations, and the attributes of these elements and relations in a workflow system. WfMC presented a Basic Process Definition Meta- model (BPDM) (Hollingsworth, 1995), as shown in Figure l.

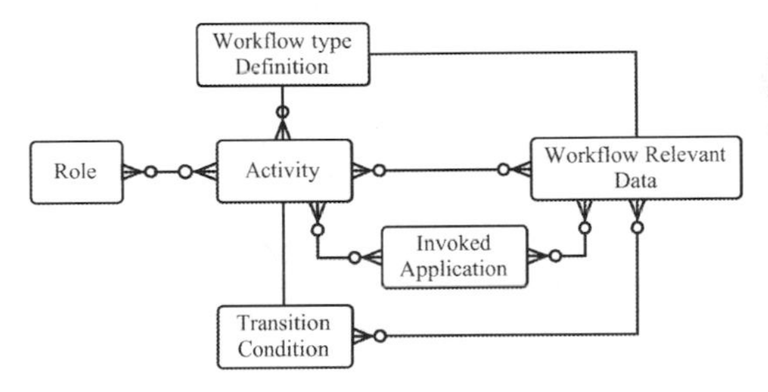

Fig. 1. Workflow process meta-model from WfMC

WfMC's BPDM statically defines the objects and their relations used in process definition stage, but, it is lack of detailed description on information model, and ignores the dynamic description of conceptions used in runtime stage. It is commonly noticed that their behavioral relations affect the relations of activities. Unfortunately, the BPDM is hard to provide a detailed description of data relations between the basic elements, making it difficult to formalize the elements and relations in a workflow system.

2.2 UML

There are several languages and notations for business process modeling, and UML (Unified Modeling Language), which is maintained by an industry consortium, namely the Object Modeling Group (OMG), is considered a main means. The UML is the result of the consolidation of several independently developed modeling languages (*e.g.*, Object Modeling Technique (OMT) and Object Oriented Software Engineering (OOSE)) from the domain of software engineering.

UML 2.0 comprehends 14 different diagram types (OMG, 2007). On the uppermost level we can discern diagrams to describe *structural aspects* of software systems and *behavioral aspects*. A well-known example for the former type is *class diagrams* that are commonly used to describe data structures in object oriented programming languages. Activity diagrams belong to the latter type of diagrams. We chose UML 2.0 version in the paper at hand because its activity diagram improves the business process representation more flexibly comparing to the previous versions. Latterly we will show how these diagrams can be extended to enable expressing geospatial related modeling.

Activity diagrams based on UML is quite suitable to describe workflows, *i.e.*, to represent a set of activities and the relations between these activities (Decker, 2009). UML formulates a series of grammar, rules, and visual symbols to express activity diagrams, making them easy to understand, share, and exchange.

2.3 OGC and ISO

There are two international organizations working on geometry or geographic-related standards for decades (Brodeur et al., 2000). One is the Open Geospatial Consortium (OGC), an international voluntary consensus standards organization since 1994. In the OGC, more than 400 commercial, governmental, nonprofit and research organizations worldwide collaborate in a consensus process encouraging development and implementation of open standards for geospatial content and services, GIS data processing and data sharing.

The OGC has a close relationship with ISO/TC 211 (Geographic Information / Geomatics), which is another international-wide technical committee working in geospatial field. An obvious evidence is that volumes from the ISO 19100 series under development by this committee progressively replace the OGC abstract specification.

Particular applications may follow standards released by OGC or ISO/TC 211 as top level guide lines. To achieve the implementation, however, many have to be tailored to meet the needs of requirements, modeling languages, and programming tools. This is because most of the OGC standards depend on a generalized architecture captured in a set of documents collectively called the *Abstract Specification*, which describes a basic data model for representing geographic features. Atop the *Abstract Specification* members have developed and continue to develop a growing number of specifications, or standards to serve specific needs for interoperable location and geospatial technology, including GIS.

Despite various specific purposes of applications, however, to be compatible with international standards is still important because: 1) location capture becomes increasingly popular since more and more devices and facilities support positioning cheaply and handy, 2) location services are more accessible since commercial and governmental agents put public and authorized service ports on web much more than before.

3 UML Models Based on Geographic Locations and Geographic Information

In this section we will find that geographic locations concern with physical definitions, while relationships on/among geographic locations are more devoted to information abstraction, therefore more semantic and flexible.

3.1 Geographic Locations

General Description

Geographic location is a natural character, either used or ignored, of any activity in business process models. Location always refers to related position of an object on the Earth surface mapping to a specific coordination reference system. For the sake of various applications, there have been defined different coordination reference systems.

Geographic location could be represented as physical position and semantic description. A physical position is a mathematical expression of an object's location. The preliminary element of a physical position is POINT, which could be represented as A (x,y) if it is defined in 2-dimensional space, or as A(x,y,z) if in 3-dimensional space. A general geographic location unit, however, is a combination of POINT, deriving into LINE, POLYGON features, furthermore into complex features, which is a combined morphology of simple features list above. It is worthy to be noticed that any geographic location should be associated with a specific coordination reference system, although different coordination reference systems could be transformed to each other. That is to say for a specific application scenario, the coordination reference system serving for all geographic locations must keep consistent.

Comparing to the mathematic, precise definition of a geographic object, human prefers to name a spatial object in natural languages. This is called semantic description. When we locate a point at (N 047° 19".416, E 05° 1".813), for instance, we would rather to name it as *Hôtel Le Jura, 14 Avenue Foch Dijon, 21000 France*. Since semantic description is quite related with particular language systems, it is closely depends on so called Geocoding mechanisms. A bidirectional channel built between physical position and semantic description is usually called Geocoding parsing.

Comparing to coordination reference systems, another important variable used to describe geographic location is *SCALE*. Scale is defined as a global parameter to express how fine a geographic object is described by a geometric feature. For example, if city of Beijing is defined in the scale of country level, it may be expressed as a POINT object; while in the scale of city, it may be expressed as a POLYGON consisting of neighborhoods, streets, urban facilities, and etc. That is to say the feature definition of a geographic object could vary in terms of scales.

UML Class Model for Geographic Locations

According to the description on geographic locations above, an UML class model could be derived as in Figure 2.

Figure 2 is tailored from OGC and ISO standards. For the simplicity of class models, we treat all geographic locations as simple geospatial features, meaning only Point, Line, and Polygon are considered for UML models. This simplicity is reasonable since in practical modeling for geospatial locations used in process management, users (or managers) prefer to use semantic description for locations, rather than physical ones. In semantic descriptions, a geographic object usually corresponds to a simple geospatial feature (*i.e.* one of point, line, and polygon). For example, if one says Beijing, it means either a point meaning the capital of China, or a polygon with boundary and area as attributes. In this case, it has nothing to do with how the city of Beijing is composited by streets, neighborhoods, and other features.

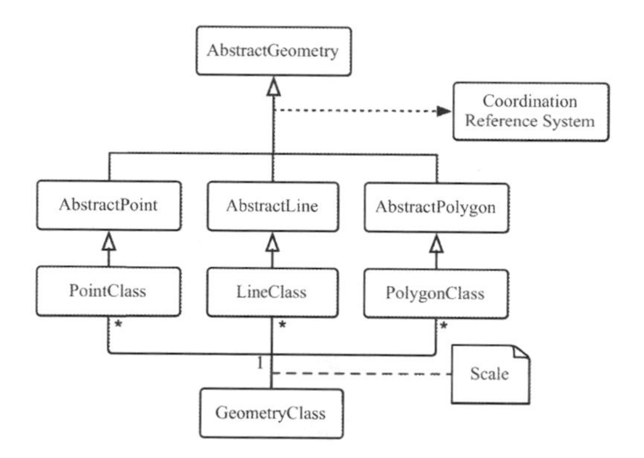

Fig. 2. UML class model for geographic locations

3.2 Geographic Information

General Description

Geographic information describes semantic relationship hidden inside geographic objects, or between two more objects. The basic categories of geographic information include measurement, sequential, and topological relationships (Theobald, 2001). All relationships could be bound together to model the scenario of real world. Since sequential relationship is less important in our case, we ignore it in this paper.

Measurement relationship depicts quantitative characteristics of an object. Typical measurement indices may reflect geometric features of objects in value, e.g., the length of a line, area of a polygon, Cartesian distance mapped to a specific coordination reference system. It also may be used to represent some natural indices such as slop of terrain, roughness of soil lands, etc. or, human geography indices such as human density of neighborhood, industrial agglomeration of an IT park, etc.

Topological relationship (*i.e* topology) expresses the spatial relationships between connecting or adjacent geospatial features (*i.e* points, lines and polygons). Topological information is useful for detecting conflicts between geospatial features (e.g. two lines in a roads vector layer that do not meet perfectly at an intersection) (Pullar and Egenhofer, 1988). More importantly, topology is necessary for carrying out some types of spatial analysis, such as network analysis, buffer calculating, and etc.

UML Class Model for Geospatial Relationship

There are many ways to model geospatial relationships with respect to different goals. ISO19107 defines a topological model based on GML (Lake et al., 2004); while OGC Topic 8 deals with relationships between features (OGC, 1999). Some applications, for example, IntesaGIS project (Amadio et al., 2004), defines the general structure and the content of a "core" geographic database in Italy consistently with the goals of the European INSPIRE project.

In this paper, we plan to define a set of UML class models to represent geospatial relationships, which is latterly used to depict activity diagrams. To this end, first we have to analyze what functionalities both measurement and topological relationships will provide, and therefore what geographic information may be derived from these relationships.

Tailored from OGC Topic 8, measurement relationships commonly used in geospatial constraints for process modeling are summarized in Table 1.

Table 1. Measurement relationship used in process modeling

Name	Description	Features involved
getCoordination	Return coordination of a point	point
getLength	Return length of a line	line
getArea	Return area of a polygon	polygon
getPerimeter	Return perimeter of a polygon	polygon

Note: a coordination reference system should be associated

Table 2. Topological relationship used in process modeling

Name	Description	Features involved
isWithin	Returns true if features are within the specified distance of one another.	(point, polygon); (line, polygon); (point, line)
isEquals	Returns true if the given features are "spatially equal".	(point, point); (line, line); (polygon, polygon)
isDisjoint	Returns true if the features are "spatially disjoint".	(point, point); (point, line); (line, line); (line, polygon); (polygon, polygon)
isIntersects	Returns true if the features are "spatially intersect".	(polygon, polygon) — overlap inside
isTouches	Returns true if the features are "spatially touch".	(line, polygon); (polygon, polygon)
isCrosses	Returns true if the features are "spatially cross".	(line, line); (line, polygon)
isOverlaps	Returns true if the features are "spatially overlap".	(polygon, polygon)
isContains	Returns true if feature A "spatially contains" feature B	(polygon, polygon) — island situation

Tailored from OGC Topic 8 and ISO19107, topological relationships commonly used in geospatial constraints for process modeling are summarized in Table 2. In turn, the topological relationships can also be extended to meet specific needs from practical applications if relationships described in Table 2 are not fulfilled for some other cases.

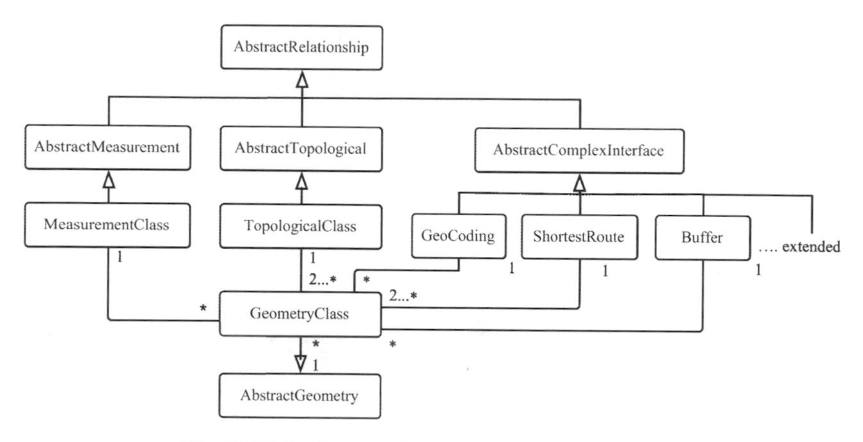

Fig. 3. UML class models of geospatial relationships

Figure 3 represents the class models of relationships analyzed above for geospatial relationships. One may notice that there is the third abstract class called *Abstract-ComplexRelationship*, which models some complex relations between two or two more geospatial objects; *Geocoding* class will do geocoding parse from physical location to semantic description, and vice versa; *ShortRoute* class will calculate the shortest route between two points with constraints such as road networks; While *Buffer* class will create a buffer polygon seeding by input features such as a point, a line, or a polygon and buffer radius.

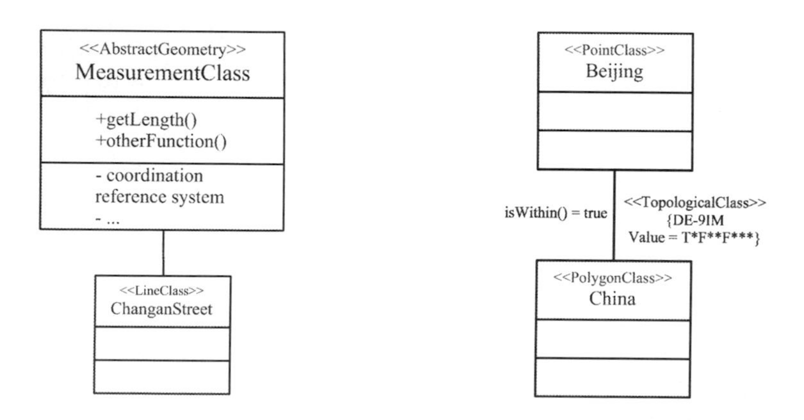

where *TopologicalClass* is declared as a stereotype extension, the tag value is defined by OGC *intersectionMatrix*, and isWithin is a constraint here, which is labeled as true.

(*a*) (*b*)

Fig. 4. Example tructure of relationship models

For *MeasurementClass*, it is easy to implement as in Figure 4*a*. The functions are defined as in Table 1 associated with some parameters (attributes). For *TopologicalClass*, there are several ways to implement it. One of the easiest way is to use DE-9IM (Dimensionally Extended nine-Intersection Model), which uses a 3x3 intersection matrix to assemble a mask code string. This mask code string could be used to interpret the 8 topological relationships in Table 2. A typical solution can be found in OGC SPEC s2.1.1.2 // s2.1.13.3 as a *Relate( )* method. Therefore, *TopologicalClass* can be constructed as shown in Figure 4*b* by a stereotype extension with *Tag* and *Value*.

4 Extended UML Activity Diagrams for Geospatial Constrained Process

In this section, a series of extended UML activity diagrams are described based on process meta-model mentioned in section 2. These diagrams will be used to answer questions such as: 1) what are geospatial constraints? 2) what are relationships between geospatial constraints and activities? And 3) how will geospatial constraints act on activities?

From the process meta-model we find that an activity in a process is pre-defined by the workflow schema, which is treated as a static definition before an instance of the process is executed. The activity has to be associated with a role (or user), and driven by workflow data assigned to it. Optionally, the activity may be triggered by some transforming conditions, which enforce this activity towards to the next one, or be influenced by some involved application, which could be an inner invocation or an external one. Therefore, elements acting on an activity could be divided into 2 types: one is the static (for example, the pre-defined workflow schema), and the other is dynamic, which is triggered only when an instance of the workflow is running.

To narrow the topic on this paper, we only consider how geospatial constraints will act on activities. The next section will give a detailed analysis on the actions.

4.1 Geospatial Constraints

A *geospatial constraint* is a statement about the geographic information on which one or more activities of a workflow schema or instance have to be performed. Geospatial constraints are somehow restraint mechanisms acting on activities by using attributes of a location or relationship between locations.

Observing the activity model of a process, there exist *static* and *dynamic* constraints (Decker, 2010). Static constraints are pre-defined for the workflow schema *before* the runtime of the individual workflow instances. This implies that 1) these constraints usually act on a single activity; and 2) these constraints are enforced for all workflow instances that are created from that schema. Dynamic constraints, however, are defined *during the runtime* of a workflow instance, and are only valid for that instance. This implies that dynamic constraints always involved with the activities on which at least one has mobile locations.

That is to say activities also have either static or dynamic (or mobile) behaviors. For a on-site repair process instance, the dispatching activity is dynamic since the worker associated with the dispatching is moving; while the repair activity is static since the location where the repair request is asked is fixed. To be more detailed, we distinguish *external* and *internal* dynamic constraints. External constraints arise when the actual locations are not calculated by the flexible workflow system itself; rather, they are defined manually by a human operator during runtime (e.g., dispatcher, manager) or they are queried from another information systems, e.g., a GIS system who provides a web-based service for public supports. When an internal constraint is defined then all the information needed by the flexible workflow system to calculate the actual location during the runtime of the workflow instance has to be specified in the workflow graph.

4.2 Actions of Geospatial Constraints

Action defines rules that how geospatial constraints will be acted on an activity, or activities. Activities are not always static. Before an instance of a process is triggered, not only static characters of activities have to be defined, but also relations between activities have to be set. While in runtime the relations could be dynamically changed. This revision on activities usually can be handled by workflow schema, if the revision has nothing to do with spatial factors. In this paper, however, we will discuss how geospatial constraints will control behaviors of activities.

Geospatial constraints only act on activities with spatial characters. It is obvious that geospatial constraints may exist against one activity, or among two, even more activities. For the former case, actions could be used to limit locations of activities to be initialized. For the latter, however, actions could be used to restrain the relations of locations where activities happen; for example, two activities have to be initialized at the same locations, or should not be initialized at the same locations.

Although some researchers analyzed that locations will perform 2 actions, *i.e.* positive satisfaction (a must-be confirm) and negative satisfaction (a must-be denial), which is an either/or case, with respect to activities' characteristics, or *acivity-to-activity* relationships, we unify the actions as one type, *i.e.* a must-be satisfaction. And with this intension, we have to transform the denial description to the *actions* on activities to the denial expression of *constraints*, which will unify the description of actions, simplifying the modeling of actions based on UML. For example, if we define two activities should not be initialized at the same locations, which is previously defined as a negative of 'the same locations', we instead define the two activities have to satisfy a topological constraint of *isDisjoint*. Since this transform is sematic based, it could be sometimes ambiguous. For example, when defining two places are of 'the-same-location' relation, we do not mean that the two locations are exactly, or mathematically equal. In perspective of semantic meaning, we imply that the two places are possibly satisfied by a topology of *isWithin*, or of *isContains*, even of *isOverlaps*.

Since the topological relationships quoted in our paper are born from OGC/ISO standards, they are comprehensive. In other words, because they are mathematically based, therefore they are strictly defined. In certain cases spatial constraints defined by some researchers previously are discrete, while our definitions on geospatial constraints are continuous in mathematical space. That is to say the former can be treated as exception of the latter.

Table 3 explains relationship that the geospatial constraints against actions on activities. In the table, the round box with shade background represents an activity with mobile geospatial character, which is dynamic, while the round box without background represents a static geospatial location, which is usually pre-defined before an instance of process is activated.

It is worthy to notice that some common control-flow patterns (such as AND, XOR, Loop, etc) have not been discussed and integrated with geospatial constraints in our paper. The reason is that we believe those common patterns explicitly control the process flow; And static and dynamic actions on activities play inner roles. The interaction between patterns above and geospatial constraints used in our paper is another topic which could be described in our latter paper; And they are quite beyond our discussion here.

Table 3. Relationship that the geospatial constraints against actions on activities

	Actions	
	on SINGLE activity	on 2+ activities
Static	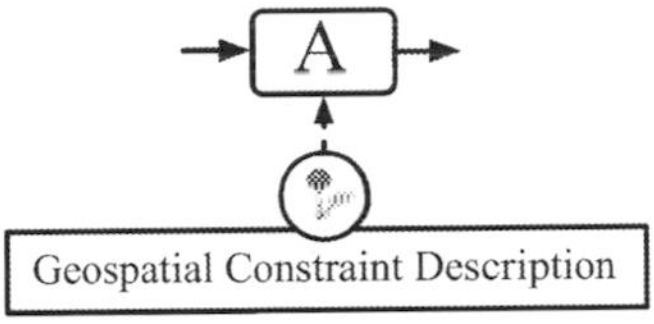 A; Geospatial Constraint Description	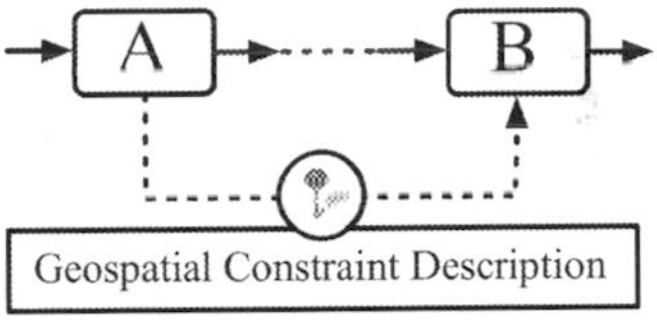 A, B; Geospatial Constraint Description
Dynamic	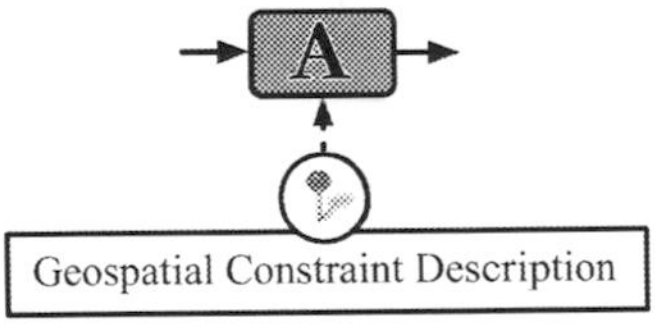 A; Geospatial Constraint Description	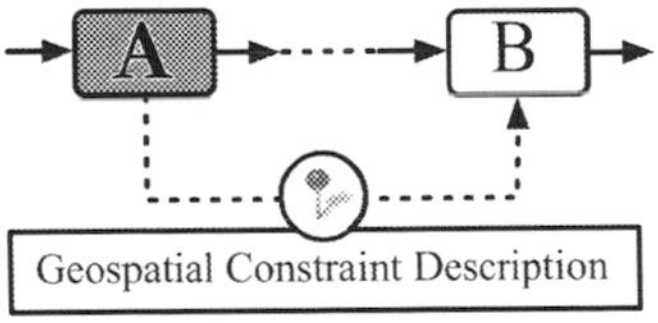 A, B; Geospatial Constraint Description

5 Case Scenario

To exemplify the application of our UML profile, a case scenario (see Figure 5) is referenced from Decker (2010). This workflow shows a company's work that provides technical maintenance services and employs many technicians that are sent to the customers' premises to perform on-site repair works of technical components. The reason why we chose this case is that we try to compare our solution with Decker's. The workflow is quite clear to understand. To emphasize some key points in our solution, we made a little revise on workflow rules:

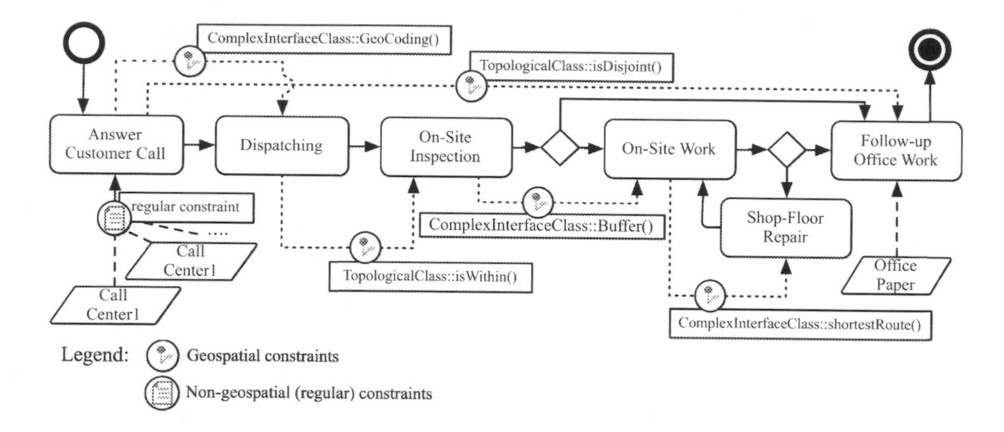

Fig. 5. Case workflow with different kinds of geospatial constraints

- ***Activities:***
  - Answer customer call: To receive customer's request and prepare to dispatch an inspector to the site of trouble;
  - Dispatching: To send an Inspector who is mobile to the site of trouble and prepare to do on-site inspection;
  - On site inspection: Inspector does on-site detect; if no trouble is found, jump the 'office work' activity; if found, then send a mobile request to all repair teams within 150 meters around his position;
  - On site work: The mobile repair team makes an on-site repair; if failed then send broken components to a nearest repair shop; this could be iterations till the trouble is removed;
  - Shop floor repair: To repair broken components;
  - Follow-up office work: To document the repair process, and make office papers for archives.
- ***Data:***
  - Call center locations: A list of locations of call centers the company deployed for collecting customer's requests.
  - Office papers: Document describing a instance of a process.
- ***Geospatial constraints used in this case:***
  - Geospatial constraints are symbolized as icon plus a description associated with.
  - ShortestRoute: When answer customer calls, only an inspector nearest to the site of the repair request will be assigned.
  - Geocoding: Geocoding is used to parse the location of customer's phone call.
  - Contain: Contains relationship is used to decide if an inspector's service area covers customer's location; or to decide if customer's location is *Within* the inspector's service area, that of *TopologicalClass::Within()*.
  - Disjoint: Disjoint relationship is used to prevent the location of answer customer call and location where the follow up office work is made from the same places.
  - Buffer: Buffer is operated to find any mobile team close to the inspector's position no more than 150 meters.

- ***Non-geospatial constraints:***
  - Call center assignment: Which call center will respond to customer's phone call will be decided by program-controlled switchboard; this is only used to demonstrate that not every activity involved locations is restrained by a geospatial constraint.
- ***Roles***:
  - Call center: Call center 1 is responsible to answer customer's request, and to dispatch an inspector to the site of trouble. While call center 2 is responsible to document office work for the instance of a process. Call center 1 and 2 have to be separated geo-spatially.
  - Inspector: Inspector is responsible to detect if a trouble exists, and call a mobile repair team for on site repair.
  - Mobile repair team: Mobile repair team is responsible to do on site repair, and assemble components repaired by a repair shop.
  - Repair shop: Repair shop is responsible to repair components sent by mobile repair team.

6 UML Profile

A profile in UML provides a generic extension mechanism for customizing UML models for particular domains and platforms. Extension mechanisms allow refining standard semantics in strictly additive manner, preventing them from contradicting standard semantics. Profiles are defined using stereotypes, tag definitions, and constraints, which are applied to specific model elements, like Classes, Attributes, Operations, and Activities. A Profile is a collection of such extensions that collectively customize UML for a particular domain or platform.

Figure 6 shows how the proposed extension to UML activity diagrams fits into the meta-model of UML. For the sake of brevity only the most important partials are constructed.

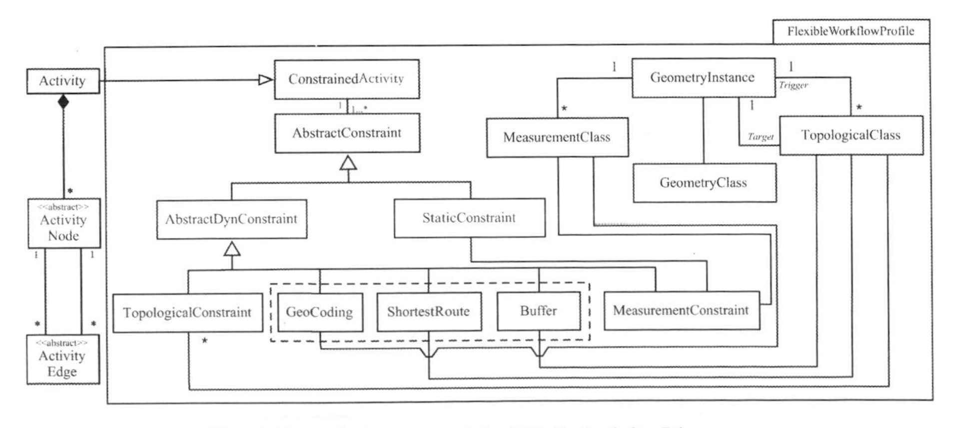

Fig. 6. Extended meta-model of UML Activity Diagrams

In Figure 6 we name the package *FlexibleWorkflowProfile* to represent the new UML profile. Also some important classes from the UML meta-model are shown outside the boundary of the package.

The profile is divided into two parts. The left part represents the geospatial constraints model, and the right the geographic information model based on geometry class models.

The main connection between the UML meta-model and our profile is the extension relationship between class *Activity* and *ConstrainedActivity*. Instances of *ConstrainedActivity* have to be used if any kind of location constraint has to be assigned to an activity. For this each *ConstrainedActivity* stands in association with at least one instance of class *AbstractConstraint*. There are two direct subclasses of *AbstractConstraint*, namely *AbstractDynConstraint* and Abstract*StaticConstraint*. *AbstractDynConstraint* has two direct subclasses: *TopologicalConstraint* and *MeasurementConstraint*.

7 Conclusions

In this paper we studied an extended process model under geospatial constraints using UML 2.0 activity diagrams. The geometry (*i.e.* point, line and polygon) models and geospatial relationship (*i.e.* measurement, sequential, and topology) models used in the paper are tailored from OGC and ISO/TC 211 standards with respect to requirements of process modeling. On the basis of these models, the paper analyzes how geospatial information restrains activities in process meta-model and how it acts on activities. With the benefits of comprehensive definitions of geospatial relationships, we unify the action as only one rule, *i.e.* a must-be satisfaction, making UML-based activity models more simplified and universal. Further, an UML profile is derived based on UML activity diagrams, and is validated by a case scenario. The contribution of this paper is marked as two points. First, the geometry models and geospatial relationship models proposed in this paper are based on abstract specification of international standards, making them better for understanding, sharing, and exchanging. Second, the geospatial constraints relied on geospatial relationship models adapt to most cases in process modeling, making them more generalizable and extendable.

Our research work has to be push deeper in the future. One is that interaction between common control-flow patterns and geospatial constraints used in this paper is an interesting topic; And based on the UML activity diagrams described in this paper at hand, we plan to implement a prototype model to validate the hypothesis proposed in this paper.

Acknowledgement. The authors would like to appreciate for the support of The National Key Technology R&D Program, China (Grant No. 2012BAH01F02).

References

Amadio, G., Cannafoglia, C., Corongiu, M., Desideri, M., Rossi, M.: IntesaGIS: The Bases for a National Spatial Data Infrastructure. In: 10th EC-GI & GIS Workshop, ESDI: The State of the Art Warsaw. Poland (2004)

Brodeur, J., Bédard, Y., Proulx, M.-J.: Modelling geographic Application databases using UML-based repositories aligned with international standards in geomatics. In: 8th ACM Symposium on GIS, Washington, D.C. (2000)

de Leoni, M., Adams, M., van der Aalst, W.M.P., ter Hofstede, A.: Visual Support for Work Assignment in Process-Aware Information Systems: Framework Formalisation and Implementation. Decision Support Systems 54(1), 341–361 (2012)

Decker, M.: Location-Aware Access Control: An Overview. In: Wireless Applications and Computing, Carvoeiro, Portugal (2009)

Decker, M.: Modelling of Mobile Workflows with UML. International Journal on Advances in Telecommunications 3(1&2), 59–71 (2010)

Decker, M., Stürzel, P., Klink, S., Oberweis, A.: Location Constraints for Mobile Workflows. In: Proceedings of the 2009 Conf. on Techniques and Applications for Mobile Commerce, TAMoCo, pp. 93–102 (2009)

Hollingsworth, D.: The Workflow Reference Model. Workflow Management Coalition (1995)

Lake, R., Burggraf, D.S., Trninic, M.: Geography Mark-up Language: Foundation for the Geo-web. John Wiley & Sons (2004)

OGC Relationships between Features. The OpenGIS Abstract Specification, p. 17 (1999)

OMG Unified Modeling Language. Object Management Group (2007)

Pullar, D., Egenhofer, M.: Towards Formal Definition of Topological Relations among Spatial Objects. In: Marble, D. (ed.) Proceedings of the Third International Symposium on Spatial Data Handling, Sydney, Australia, pp. 225–242 (1988)

Rosemann, M., Recker, J., Flender, C.: Contextualisation of Business Processes. International Journal of Business Process Integration and Management 3(1), 47–60 (2008)

Rosemann, M., Recker, J., Flender, C., Ansell, P.: Understanding Context-Awareness in Business Process Design. In: 17th Australasian Conference on Information Systems, Adelaide, Australia (2006)

Theobald, D.M.: Topology revisited: representing spatial relations. International Journal of Geographical Information Science 15(8), 689–705 (2001)

Van der Aalst, W.M.: Process Mining. Springer, Heidelberg (2011)

Van der Aalst, W.M.: Workflow Verification: Finding Control-Flow Errors Using Pebi-Net-Based Techniques. In: Business Process Management, pp. 161–183 (2000)

Zhang, L., Zhao, J., Jia, W., Liu, B.: Location-aware Workflow Modeling and Soundness Verification Method Based on Petri net. Computer Integrated Manufacturing Systems 18(008), 1747–1756 (2012)

Generating Synthetic Process Model Collections with Properties of Labeled Real-Life Models

Zhiqiang Yan[1,2], Remco Dijkman[3], and Paul Grefen[3]

[1] Capital University of Economics and Business, Beijing, China
[2] Tsinghua University, Beijing, China
[3] Eindhoven University of Technology, Eindhoven, The Netherlands
zhiqiang.yan.1983@gmail.com, {r.m.dijkman,p.w.p.j.grefen}@tue.nl

Abstract. Nowadays, business process management plays an important role in the management of organizations. More and more organizations describe their operations as business processes. It is common for organizations to have collections of thousands of business processes, but due to privacy reasons, these collections are often not, or only partially, available to researchers. However, to test the scalability of techniques it is necessary to have collections the size of which is several orders of magnitude larger than the collections that are available. Therefore, this paper proposes a technique to generate such collections of process models based on the properties of real-life collections. Where existing techniques focus on the structure of the process models, the technique proposed in this paper also generates brief task labels. This allows BPM experimentation with process collections that have laboratory-set quantitative parameters and real-world qualitative characteristics.

1 Introduction

Nowadays, business process management technologies develop quickly in both academic and industrial fields. As a result, it is common to see collections of hundreds or even thousands of business process models. For example, the SAP reference model consists of more than 600 business process models [1], and the information department of China Mobile Communication Corporation (CMCC) maintains more than 8,000 processes in its BPM systems [3]. To manage such a large number of process models efficiently and automatically, business process model repositories [7,9] are required. These repositories provide techniques including process similarity search [10], process querying [6], process storage management [8].

Real-life business process models are the property of companies and are usually not made publicly available. Consequently, it is difficult for researchers to get a large amount of real-life process models to evaluate their techniques for managing business process models. However, such real-life collections of business process models are needed to evaluate the quality and performance of algorithms that have been developed. Examples of such algorithms are algorithms for searching through a collection of business process models to find models that

C. Ouyang and J.-Y. Jung (Eds.): AP-BPM 2014, LNBIP 181, pp. 74–88, 2014.

are similar to a given model [10], querying a collection to find models with certain properties [11], and refactoring a collection of process models [2]. To solve the disconnect between the availability of large collections of models from practice and the necessity to have such collections for evaluation purposes, synthetic collections of models are often created.

For example, in the graph database area, synthetic graphs are generated to evaluate different related techniques [4]. Similarly, in the BPM area, Hee et al. [5] propose a technique to generate synthetic models based on a given process model collection. However, this technique focuses on structure and does not consider real-life activity labels, instead labeling activities with codes to distinguish them from each other. In this paper, we propose a technique that generates process models with complex activity labels (besides preserving structural properties) based on the real-life properties of a given process model collection. This extension is necessary, because some algorithms, and especially algorithms for searching and querying business processes, rely on the properties of activity labels as well as the structure of business process models. Therefore, in order to properly evaluate such algorithms, it is necessary to generate business process models with labels that have real-life properties.

The properties that are considered include size of process models, number of common workflow patterns (sequences, splits, joins, and loops), words in labels, size of labels, and co-occurrences of words. The steps of generating process models are as follows. First, the properties of process models are abstracted from a given collection. Second, labels and workflow patterns are generated based on the properties and probabilities. Last, for a new synthetic model, a (generated) pattern is inserted each time until its size reaches the pre-selected number.

The rest of the paper is organized as follows. Section 2 presents preliminaries. Section 3 explains how properties are abstracted from a given process model collection. Section 4 presents algorithms to generate synthetic process models. Section 5 presents the experiments. Section 6 concludes this paper.

2 Business Process Graph

We define the technique on process graphs. A process graph is a graph-based representation of a process model. The benefit of using a graph-based representation is that it can be used to represent the structure of existing (graph-based) business process modeling languages. This includes advanced constructs such as boundary events, which can be represented using a typed edge. Process graphs can be transformed to models that are constructed with different business process modeling languages. As an example, Fig. 1 shows five BPMN models and their corresponding business process graphs. As shown in *graph 4* of Fig. 1, we assign each gateway node a unique label to represent its routing function, e.g., *'And-Split'* and *'Xor-Join'*.

Definition 1 (Process Graph, Pre-set, Post-set). *Let $\mathcal{L}$ be a set of labels. A process graph describes a (business) process as a tuple (N, E, λ), in which:*

- *N is the set of nodes.*
- *$E \subseteq N \times N$ is the set of edges.*
- *$\lambda : N \to \mathcal{L}$ is an injective function that maps nodes to labels.*

Let $G = (N, E, \lambda)$ be a process graph and $n \in N$ be a node: $\bullet n = \{m|(m,n) \in E\}$ is the pre-set of n, while $n\bullet = \{m|(n,m) \in E\}$ is the post-set of n.

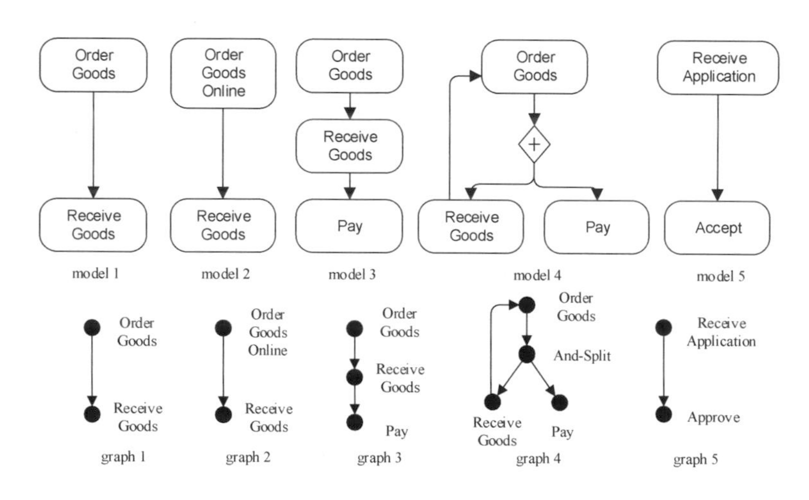

Fig. 1. A Collection of Business Process Models

We consider the most common workflow patterns: sequence, split, join, and loop as features, which are used to compose synthesis graphs later.

Definition 2 (Basic Feature). *Let $\mathcal{D}$ be a collection of process graphs and $g \in \mathcal{D}$ be a process graph. A feature f of g is a subgraph of g. The size of a feature is the number of edges it contains, denoted as $Size(f) = |E_f|$. Let max be a threshold, indicating the maximal size of a feature that is considered. The type of a feature is the structural pattern of a feature, including, denoted $Type(f) \in \{node, sequence, split, join, loop\}$. Feature f is*

- *a sequence feature of size $s-1$ consisting of nodes $\{n_1, n_2, \dots, n_s\}$, if E_f is the minimal set containing $(n_1, n_2), (n_2, n_3), \dots, (n_{s-1}, n_s)$, for $s \geq 2$. It is denoted as $n_1 \to n_2 \to \dots \to n_s$.*
- *a split feature of size s consisting of a split node n and a set of nodes $\{n_1, n_2, \dots, n_s\}$, if and only if E_f is the minimal set containing (n, n_1), $(n, n_2), \dots, (n, n_s)$, for $s \geq 2$. It is denoted as $n \to \{n_1, n_2, \dots, n_s\}$.*
- *a join feature of size s consisting of a join node n and a set of nodes $\{n_1, n_2, \dots, n_s\}$, if and only if E_f is the minimal set containing (n_1, n), $(n_2, n), \dots, (n_s, n)$, for $s \geq 2$. It is denoted as $\{n_1, n_2, \dots, n_s\} \to n$.*
- *a loop feature of size s consisting of nodes $\{n_1, n_2, n_3, \dots, n_s\}$, if E_f is the minimal set containing $(n_1, n_2), (n_2, n_3), \dots, (n_{s-1}, n_s), and (n_s, n_1)$, for $s \geq 1$. It is denoted as $n_1 \to n_2 \to \dots \to n_{s-1} \to n_s$.*

The sequence, split, join, and loop features are referred as structural features.

For example, for *graph 4* in Fig. 1, the set of basic node features consists of nodes *'Order Goods', 'Receive Goods', 'And-Split', and 'Pay'*; the set of the basic sequence features of size 1 consists of sequences *'Order Goods'→'And-Split', 'And-Split'→'Receive Goods', 'And-Split'→'Pay', and 'Receive Goods'→'Order Goods*; the basic split feature set consists of the feature with split node *'And-Split'* and the set of nodes *'Receive Goods', 'Pay'*; and the basic loop feature set consists of the loop feature with three basic nodes *'Order Goods', 'And-Split', and 'Receive Goods'.*

A feature may consist of other smaller features, which are called parent (child) features.

Definition 3 (Parent Feature, Child Feature). *If a feature f can generate feature cf by adding a single edge and at most one node, feature f is a direct parent feature of feature cf and feature cf is a direct child feature of feature f. It is denoted as $f \in DPFS(cf)$ or $cf \in DCFS(f)$, where $DPFS$ ($DCFS$) is a function that maps a feature to its direct parent (child) feature set. The parent and the child relation are the transitive closure of the direct parent and the direct child relation.*

For example, for *graph 4* in Figure 1, the direct child feature set of the node feature 'Receive Goods' is the set consisting of sequence features *'And-Split'→'Receive Goods'* and *'Receive Goods'→'Order Goods'.*

3 Properties of Business Process Model Collections

This section presents properties that are discovered from business process model collections and that are relevant to evaluate the process querying technique. The two most common aspects of business process models [9], are considered, i.e., activity and control flow. For activities, properties of node (label) features are discovered. For control flows, properties of structural features of process models are discovered.

3.1 Label Properties

Labels consist of words and words in labels of a given process model collection are composed to form synthetic labels in this section. To generate synthetic labels with similar properties as labels in a collection, we consider two types of label properties regarding words, i.e., the occurrence of a word and the co-occurrence of two words. The former one indicates the frequency and probability of a word appearing in a synthetic label and the later one indicates the frequency and probability of two words appearing in a synthetic label or labels of two connected nodes.

Lower case versions of the words are used. Stop-words, e.g., 'a', 'an', 'the', 'one', ..., and gateway labels, e.g., 'and-split', are not considered. Definition 4 presents the word set of a process model collection.

Definition 4 (Word Set, Label Size).
Let $\mathcal{D}$ be a collection of process graphs with disjoint sets of nodes and let $\mathcal{L}$ be the label set. The function $\omega(l)$ maps a label l to the set of words that appear in l.

The word set $\mathcal{W}$ of the collection $\mathcal{D}$ is the set of words appear in $\mathcal{L}$. Formally, $\mathcal{W} = \{w | w \in \omega(l) \wedge l \in \mathcal{L}\}$.

The size of a label l is the number of words in l, i.e., $|\omega(l)|$.

For example, the word set of the collection in Figure 1 is $\{order, goods, receive, online, pay, application, approve\}$.

To generate a label, it is necessary to know which word is in the label. Therefore, the probability of word occurrence is required. For a word in the word set of a collection, the frequency and probability of its occurrence are defined in as follows.

Definition 5 (Frequency of Word Occurrence, Probability of Word Occurrence). *Let $\mathcal{D}$ be a collection of process graphs with disjoint sets of nodes, let $\mathcal{N}$ be the node set of $\mathcal{D}$, and $\mathcal{W}$ be the word set of $\mathcal{D}$.*

The frequency of the occurrence of a word w, denoted as $FWO(w)$, is the number of nodes in the collection that contain the word w in their labels. Formally, $FWO(w) = |\{n \in \mathcal{N} | w \in \omega(\lambda(n))\}|$.

The probability of the occurrence of a word w is the frequency of the occurrence of w divided by the frequency of the occurrence of all words. Formally, $PWO(w) = \frac{FWO(w)}{\sum_{w_1 \in \mathcal{W}} FWO(w_1)}$.

For example, in Figure 1, the frequency of the word occurrence of *'goods'* is 8, and its probability is 8/22=0.36.

To generate a label, it is also necessary to know which words can occur in the same label or labels of two connected nodes. Therefore, three types of word co-occurrence are considered as defined in Definition 6, i.e., word co-occurrence, pre-word co-occurrence, and post-word co-occurrences. Gateway nodes are not considered for pre-word (post-word) co-occurrence. For a node n, if a node $n_1 \in \bullet n$ $(n\bullet)$ is a gateway node, nodes in $\bullet n_1$ $(n_1\bullet)$ of the gateway nodes are considered instead of n_1. For example, in *graph 4* of Figure 1, the post-word co-occurrence for words in the node *'order goods'*, nodes *'receive goods'* and *'pay'* are considered instead of the gateway node *'and-split'*.

The frequencies of three types of word co-occurrence are defined as follows.

Definition 6 (Frequency of Word Co-Occurrence). *Let $\mathcal{D}$ be a collection of process graphs with disjoint sets of nodes, $\mathcal{N}$ be the node set of $\mathcal{D}$, and $\omega(l)$ is the function that maps a label l to the set of words that appear in l.*

- *Frequency of Word Co-Occurrence (FWCO): Given a word w and another word w_1 $(w_1 \neq w)$, the frequency of the word co-occurrence of w and w_1 is the number of nodes of the collection $\mathcal{D}$ that contain both w and w_1. Formally, $FWCO(w, w_1) = |\{n \in \mathcal{N} | w, w_1 \in \omega(\lambda(n))\}|$. We say that word w_1 co-occurs in the same node label with w if $FWCO(w, w_1) > 0$.*

- *Frequency of Pre-Word Co-Occurrence ($FWCO_{pre}$): Given two words w and w_1, the frequency of the pre-word co-occurrence of w_1 with respect to w is the number of process fragments (a sequence of two nodes) satisfying that w appears in the label of a node n, w_1 appears in the label of a node n_1, and there exists a process graph g, in which n_1 is in the pre-set of n. Formally, $FWCO_{pre}(w, w_1) = |\{n_1 \in \mathcal{N} | g \in \mathcal{D} \wedge n, n_1 \in N_g \wedge n_1 \in \bullet n \wedge w \in \omega(\lambda(n)) \wedge w_1 \in \omega(\lambda(n_1))\}|$. We say that word w_1 co-occurs with w in a pre-set node label if $FWCO_{pre}(w, w_1) > 0$.*
- *Frequency of Post-Word Co-Occurrence ($FWCO_{post}$): Given two word w and w_1, the frequency of the post-word co-occurrence of w_1 with respect to w is the number of process fragments (a sequence of two nodes) satisfying that w appears in the label of a node n, w_1 appears in the label of a node n_1, and there exists a process graph g, in which n_1 is in the post-set of n. Formally, $FWCO_{post}(w, w_1) = |\{n_1 \in \mathcal{N} | g \in \mathcal{D} \wedge n, n_1 \in N_g \wedge n_1) \in n \bullet \wedge w \in \omega(\lambda(n)) \wedge w_1 \in \omega(\lambda(n_1))\}|$. We say that word w_1 co-occurs with w in a post-set node label if $FWCO_{post}(w, w_1) > 0$.*

For example, with respect to the word *'receive'* in the collection in Figure 1, the frequency of the word co-occurrence for *'goods'* is 4; the frequency of pre-word co-occurrence for *'goods'* is 4; the frequency of post-word co-occurrence for *'goods'* is 1.

The probabilities of the three types of word co-occurrences are defined in Definition 7.

Definition 7 (Probabilities of Word Co-Occurrences). *Let $\mathcal{D}$ be a collection of process graphs with disjoint sets of nodes, $\mathcal{N}$ be the node set of $\mathcal{D}$, $\mathcal{W}$ be the word set of $\mathcal{D}$, and $\omega(l)$ is the function that maps a label l to the set of words that appear in l. Three Types of word co-occurrence probabilities are defined as follows.*

- *Probability of Word Co-Occurrences ($PWCO$): Given a word w and another word w_1, the probability of word co-occurrences of w and w_1 is the frequency of w_1 co-occurring with w divided by the frequency of all words co-occurring with w. Formally,*

$$PWCO(w, w_1) = \frac{FWCO(w, w_1)}{\sum_{w_2 \in \mathcal{W} \wedge w_2 \neq w} FWCO(w, w_2)}. \tag{1}$$

- *Probability of Pre Word Co-Occurrences ($PWCO_{pre}$): Given two words w and w_1, the probability of the co-occurrences of w_1 with respect to w is the frequency of w_1 co-occurring with w in a pre-set node label divided by the frequency of all words co-occurring with w in a pre-set node label. Formally,*

$$PWCO_{pre}(w, w_1) = \frac{FWCO_{pre}(w, w_1)}{\sum_{w_2 \in \mathcal{W}} FWCO_{pre}(w, w_2)}. \tag{2}$$

- *Probability of Post Word Co-Occurrences ($PWCO_{post}$): Given two words w and w_1, the probability of the co-occurrences of w_1 with respect to w is the* *frequency of w_1 co-occurring with w in a post-set node label divided by the frequency of all words co-occurring with w in a post-set node label. Formally,*

$$PWCO_{post}(w, w_1) = \frac{FWCO_{post}(w, w_1)}{\sum_{w_2 \in \mathcal{W}} FWCO_{post}(w, w_2)}. \quad (3)$$

For example, with respect to the word *'receive'* in the collection in Figure 1, the probability of word co-occurrence for *'goods'* is 4/5=0.80; the probability of pre-word co-occurrence for *'goods'* is 4/9=0.44; the probability of post-word co-occurrence for *'goods'* is 1/3=0.33.

3.2 Structural Properties

The structure of a process model can be described in terms of a set of common patterns, which are called features in this paper. In this section, properties of these features are considered as structural properties, e.g., feature type and feature size as defined in Definition 2. Furthermore, the composition rules are abstracted, which indicate how features can be composed to form a process graph.

As defined in Definition 2, four types of structural features are considered, i.e., sequence, split, join, and loop. Process graphs consists of compositions of these features. For example, *graph 6* of Figure 2 consists of a sequence feature $a \to b \to c$, a split feature $b \to \{c, d, e, f, g\}$; a join pattern $\{c, d\} \to h$, a loop feature $f \to k$, etc.

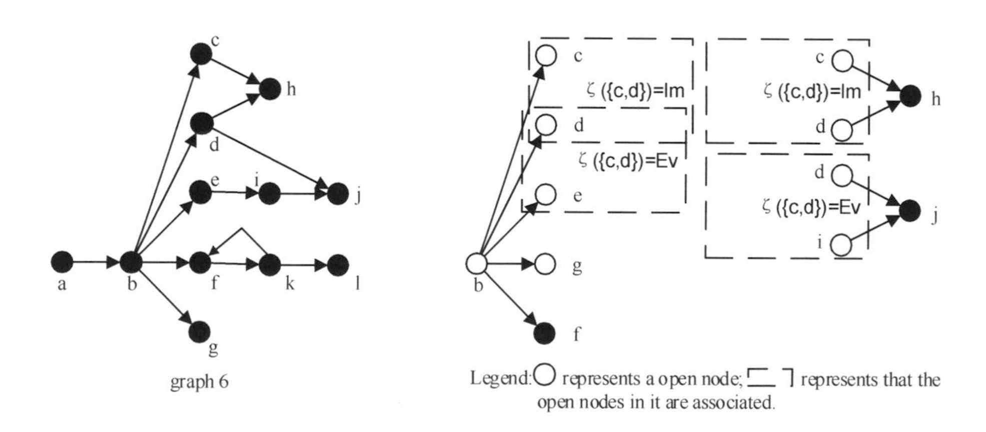

Fig. 2. A Business Process Graph and Associations of Open Nodes

When splitting a process graph into features, a feature can be a parent of different features (parent and child features are defined in Definition 3). For example, in *graph 6* of Figure 2, sequence feature $b \to c$ is a parent of sequence features $a \to b \to c \to h$, $b \to c \to h$, etc. This affects the probability of the occurrence of a certain type of feature. For example, sequence features are counted too many times, since all structural features consists of sequence features of size 1. Therefore, only local maximal features are considered, which are features without child features, as defined in Definition 8. For example, sequence feature $b \rightarrow c$ is not considered any more; while sequence feature $a \rightarrow b \rightarrow c \rightarrow h$ is considered.

Definition 8 (Local Maximal Feature). *Let $g = (N, E, \lambda)$ be a business process graph. Let f, a subgraph of g, be a feature. The feature is a local maximal feature of g, denoted as $f \in LMF(g)$, if and only if there is not another feature f_1 of g, such that f is a parent feature of f_1.*

In a process graph, a node can be in different local maximal features and have different pre-sets or post-sets in these features. These nodes are called open nodes. In Section 4 when generating process graphs, we use these nodes as the points to extend a process graph and create a large synthetic graph. An open node is defined as follows.

Definition 9 (Open Node, Closed Node). *Let g be a process graph and f be a feature of g. A node n of the feature f is an open node for that feature, if and only if the feature does not contain all of the nodes in the pre-set or post-set of n. Formally, $\forall n \in N_f : n \in ON(f) \iff \exists((\bullet n \cup n\bullet) - N_f) \neq \emptyset$. A node n in the feature is an pre-open (post-open) node for the feature, if and only if the feature does not contain all of the nodes in the pre-set (post-set) of n, denoted as $n \in ON_{pre}(f)$ ($n \in ON_{post}(f)$). A node is a closed node if it is not an open node.*

For example, in *graph 6* of Figure 2, node d is a post-open node for split feature $b \rightarrow \{c, d, e, f, g\}$, and node d is both a pre-open and post-open node for join feature $\{c, d\} \rightarrow h$. When generating a synthetic graph, given $b \rightarrow \{c, d, e, f, g\}$ is already in the synthetic graph, node d can be used to extend the graph with another feature that has a pre-open node. More details about extending a synthetic graph is given in Section 4.

For now, we know which node in a feature is an open node that can be used to extend another feature. However, for a split or join feature, a subset of open nodes can be associated to extend features. The split or join association of open nodes is to support a structure like that after a split some of the branches join together immediately or eventually. For example, in Figure 2, after the split node b, the branches of node c and node d join together immediately at node h; the branches of node e and f join together eventually at node j. The split or join association of open nodes is defined in Definition 10.

Definition 10 (Split (Join) Association of Open Nodes). *Let g be a business process graph and f be a split feature. Let the node $n \in N_f$ be the split node and let $N' = ON_{post}(f)$ be the post-open node set for f. Let $N'' \subseteq N'$ ($|N''| > 1$) be a subset of the post-open nodes. If there exists a node $n_2 \in N_g$, such that for each $n_1 \in N''$, $(n_1, n_2) \in E_g$, we say that f has an immediate join, denoted as $\zeta(N'') = Im$; otherwise if there exists a node $n_2 \in N_g$, such that for* *each* $n_1 \in N''$, *there exists a sequence feature in* g, $n_1 \to \ldots \to n_2$, *we say that* f *has an eventual join, denoted as* $\zeta(N'') = Ev$. Im *and* Ev *are the type of a split (join) association of open nodes. The join association of open nodes is defined similarly.*

For example, on the right side of Figure 2, it shows some split and join features of *graph 6*, the split feature $b \to \{c, d, e, f, g\}$ has an immediate join $\{c, d\}$ and an eventual join $\{d, e\}$; while the join feature $\{c, d\} \to h$ has an immediate split $\{c, d\}$ and the join feature $\{d, i\} \to j$ has an eventual split $\{d, i\}$. For an immediate join (split), a join (split) feature is inserted to extend the process graph; for an eventual join (split), a join (split) feature and several sequence features are inserted to extend the process graph.

We know that a feature can be inserted into a process graph if they both have open nodes or split (join) associations of open nodes. However, the definitions of a feature and a process graph do not contain items to indicate which nodes are open nodes or split (join) associations of open nodes. Therefore, we define a component based on a feature of a process graph to record this information. We need this information, because in next section components are composed to form a (partial) synthetic process graphs, and open nodes or split (join) associations of open nodes indicate how to compose components. A component is defined as follows.

Definition 11 (Component). *Let* g *be a business process graph and let* f *be a local maximal feature of* g. *The component of* f, *denoted as* $c = Comp(f, g)$, *is a tuple* $(N, E, \lambda, preON, postON, \zeta)$, *in which:*

- $N = N_f$ *is the set of nodes.*
- $E \subseteq N \times N$ *is the set of edges, where* $E = E_f$.
- λ *is a function that maps each node in* N *to an empty label.*
- $preON = \{n \in N_f | (\bullet n - N_f) \neq \emptyset\}$ *is the pre-open node set.*
- $postON = \{n \in N_f | (n \bullet - N_f) \neq \emptyset\}$ *is the post-open node set.*
- ζ *is a function that maps a subset of the pre-open (post-open) node set to an immediate or eventual join (split), as defined in Definition 10.*

The size and type of a component are the same as the size and type of the feature it derives from, i.e., $Size(s) = Size(f)$ *and* $Type(s) = Type(f)$.

For example, $(\{a, b, g\}, \{(a, b), (b, g)\}, \lambda, \emptyset, \{b\}, \zeta)$ is a sequence component of *graph 6* (labels are used to identify nodes here). Definition 12 presents how to abstract all components from a process graph or a collection of process graphs.

Definition 12 (Component Set). *Let* $\mathcal{D}$ *be a collection of process graphs. The component set of* $\mathcal{D}$ *consists of the components of all local maximal of process graphs in* $\mathcal{D}$. *Formally,* $C = \{Comp(f, g) | g \in \mathcal{D} \wedge f \in \Gamma(g) \cap LMT(g)\}$.

We say two components are equivalent if there is a mapping between two components, as defined in Definition 13.

Definition 13 (Component Equivalence). *Let* $c_1 = (N_1, E_1, \lambda, preON_1, postON_1, \zeta)$ *and* $c_2 = (N_2, E_2, \lambda, preON_2, postON_2, \zeta)$ *be two components. Components* c_1 *and* c_2 *are equivalent, denoted as* $c_1 = c_2$*, if and only if there exists a one-to-one mapping* $M : N_1 \to N_2$*, such that*

- $\forall n \in preON_1$: $M(n) \in preON_2$; $\forall n \in postON_1$: $M(n) \in postON_2$; $\forall n \in (N_1 - preON_1 - postON_1)$: $M(n) \in (N_2 - preON_2 - postON_2)$;
- $\forall (n, m) \in E_1$: $(M(n), M(m)) \in E_2$;
- *if* s_1 *and* s_2 *are split (join) components,* $\forall sN_1 \in postON_1(preON_1)$: $\zeta_1(sN_1) = \zeta_2(\{M(n_1) | n_1 \in sN_1\})$.

For example, in Figure 2, components $\{c, d\} \to h$ and $\{d, i\} \to j$ are not equivalent, because $\zeta(\{c, d\}) \neq \zeta(\{d, i\})$. Based on the component equivalence, the frequency and probability of component occurrence are defined as follows.

Definition 14 (Frequency of Component Occurrence, Probability of Component Occurrence). *Let* $\mathcal{D}$ *be a collection of process graphs and* C *be the component set of* $\mathcal{D}$*. Let* $c \in C$ *be a component. The frequency of the occurrence of c is the number of local maximal features in* $\mathcal{D}$ *having the equivalent component with c, denoted as* $FCO(c, C) = |\{f \in \Gamma(g) | g \in \mathcal{D} \wedge f \in LMF(g) \wedge Comp(f, g) = c\}|$*; the probability of its occurrence is the fraction between the frequency of its occurrence and the frequency of the occurrence of all components, denoted as* $PCO(c, C) = \frac{FCO(c,C)}{\sum_{c_1 \in C} FCO(c_1, C)}$.

4 Generating Synthetic Process Model Collections

This section presents the algorithm to generate synthetic process models based on the properties defined in the previous section. The algorithm consists of two steps. It first generates node labels and it then generates a synthetic graph by inserting components into the graph and labeling component nodes.

Synthetic labels can be generated based on the probabilities of word occurrence and word co-occurrence, which is defined in Definition 15.

Definition 15 (Synthetic Label). *Let* $\mathcal{D}$ *be a collection of process graphs with disjoint sets of nodes,* $\mathcal{N}$ *be the node set of* $\mathcal{D}$*,* $\mathcal{W}$ *be the word set of* $\mathcal{D}$*, and* $\omega(l)$ *be the function that maps a label* l *to the set of words that appear in* l.

A synthetic label of size s *consists of a word,* $w \in \mathcal{W}$*, and a set of* $s-1$ *words,* $W = \{w_1, w_2, ..., w_{s-1}\} \subset \mathcal{W}$ $(w \notin W)$*, which co-occur with* w.

The probability of the size s *is* $\frac{|\{n \in \mathcal{N} | \omega(\lambda(n)) = s\}|}{|\mathcal{N}|}$*. The probability of selecting a word* w *is* $PWO(w)$*, according to Definition 5. The probability of selecting a word* $w_i \in W$ *is* $PWCO(w, w_i)$*, according to Definition 6.*

A synthetic node is labeled by selecting a synthetic label from a set of synthetic labels. The selection consists of two steps. Firstly, a word is selected based on the probabilities of pre-word and post-word co-occurrence. Second, a synthetic label is selected from the subset of synthetic labels containing the selected word. The probability of selecting a label is defined as follows.

Definition 16 (Probability of Label Selection).
Let $\mathcal{SL}$ a set of synthetic labels generated according to Definition 15, let sl be a synthetic, $sl \in \mathcal{SL}$, and let $\mathcal{W} = \{w | w \in \omega(sl) \wedge sl \in \mathcal{SL}\}$ be the word set of $\mathcal{SL}$. Let c be a component and let n be a node of c, $n \in N_c$.

The probability of selecting a word w is normally the frequency of words in the pre-set/post-set of n co-occurs with w in a pre-set/post-set node label divided by the frequency of words in the pre-set/post-set of n co-occurs with any word in a pre-set/post-set node label; however, if the denominator is 0, the probability of selecting w is the probability of the occurrence of w.

$$PLS(w) = \begin{cases} \frac{\sum_{w' \in W_{post}} FWCO_{pre}(w',w) + \sum_{w' \in W_{pre}} FWCO_{post}(w',w)}{\sum_{w'' \in \mathcal{W}} (\sum_{w' \in W_{post}} FWCO_{pre}(w',w'') + \sum_{w' \in W_{pre}} FWCO_{post}(w',w''))}, \\ \text{if } \sum_{w'' \in \mathcal{W}} (\sum_{w' \in W_{post}} FWCO_{pre}(w',w'') \\ \quad + \sum_{w' \in W_{pre}} FWCO_{post}(w',w'')) \neq 0; \\ \\ PWO(w), \text{ otherwise.} \end{cases}$$

where $W_{post} = \{w | \forall n_1 \in n \bullet \wedge w \in \omega(\lambda(n_1))\}$ and $W_{pre} = \{w | \forall n_1 \in \bullet n \wedge w \in \omega(\lambda(n_1))\}$.

The probability of selecting a label sl from $\mathcal{SL}$ based on w is one divided by the number of labels that contains the word w, i.e., $P_l(w, sl) = \frac{1}{|\{sl_1 \in \mathcal{SL} | w \in \omega(sl_1)\}|}$.

Overall, the probability of selecting a synthetic label sl from $\mathcal{SL}$ is $P(sl) = \sum_{w \in \omega(sl)} (P_w(w) \times P_l(w, sl))$.

To generate a synthetic graph, the size of the synthetic graph is required, of which the probability is defined as follows.

Definition 17 (Probability of Graph Size). *Let $\mathcal{D}$ be a collection of process graphs. The probability of a graph g of size s in $\mathcal{D}$, $PGS(s, \mathcal{D})$, is that the number of graphs of size s in $\mathcal{D}$ divided by the number of all graphs in $\mathcal{D}$. Formally, $PGS(s, \mathcal{D}) = \frac{|\{g \in \mathcal{D} | |E_g| = s\}|}{|\mathcal{D}|}$.*

Algorithm 1 presents the algorithm for generating synthetic process graphs. Firstly, a set of synthetic labels are generated, according to Definition 15, which are used to label nodes in components later (line 2). Then, a set of synthetic process graphs are generated. Each graph is generated as follows. Initially a synthetic process graph *sg* contains one node *n* that is both a pre-open and post-open node ($n \in preON \cap postON$); the label of *n* is empty; The synthetic process graph does not contain any edge (line 6). Then components are inserted to extend the graph (lines 8-21).

To insert a component into a synthetic graph, an open node is randomly selected. If the open node is not in join (split) associations, a component is selected, having an open node that can be merged with the open node in the synthetic graph as explained in Algorithm 2. If the open node is in join (split) associations, one of the join (split) associations is randomly selected. Then, a component is selected, which has a join (split) association with the same number

Algorithm 1. Synthetic Process Graph Generation

```
input  : a collection of process graphs: D, an integer: size_c
output: a collection of synthetic process graphs SD
1  begin
2      SL is the set of generated synthetic labels (Definition 15);
3      SD ← ∅ ;
4      C is the component set of D (Definition 12);
5      while |SD| < size_c do
6          sg = {{sn}, ∅, ∅, {sn}, {sn}, ∅}; //sn is a newly created node.
7          select size_g ∈ {|E_g||g ∈ D} with probability PGS(size_g, D)
           (Definition 17);
8          while |E_sg| < size_g ∧ (preON_sg ∪ postON_sg) ≠ ∅ do
9              randomly select n ∈ preON_sg ∪ postON_sg;
10             if n ∈ postON_sg then
11                 if ∄N ⊆ N_sg : n ∈ N ∧ ζ_sg(N) ∈ {Im, Ev} then
12                     C_1 ← {c ∈ C|preON_c ≠ ∅};
13                     select c ∈ C_1 with probability PCO(c, C_1) (Definition 14);
14                     randomly select n_1 ∈ preON_c;
15                     sg ←mergeNode (sg, c, n, n_1, SL); //Algorithm 2.
16                 else
17                     randomly select
                       AN ∈ {N ⊆ N_sg|n ∈ N ∧ ζ_sg(N) ∈ {Im, Ev}};
18                     C_1 ← {c ∈ C|Type(c) = join ∧ AN ⊆ preON_c ∧ |AN| =
                       |AN_1| ∧ ζ_sg(AN) = ζ_c(AN_1)};
19                     select c ∈ C_1 with probability PCO(c, C_1) (Definition 14);
20                     sg ←mergeAsso (sg, c, AN, AN_1, SL);//Algorithm 3.
21             if n ∈ preON_sg then  //similar to lines 11-20.
22         SD ← SD ∪ {sg};
23     return SD;
```

of open nodes and the same type (line 19); the probability of the selection is the probability is $PCO(c, C_1)$ as explained in Definition 14. Finally, the open nodes in the join (split) associations are merged as explained in Algorithm 3.

Algorithm 2 presents the steps of merging an open node of a synthetic graph and an open node of a component. Firstly, (open) nodes in the component are inserted the (open) node set of the synthetic graph; the open nodes to be merged are not open nodes anymore. Secondly, edges connected to the open node in the component connect to the open node in the synthetic graph; other edges in the component are inserted into the edge set of the synthetic graph. Thirdly, nodes in the component expect for the open node are labeled according to Definition 16.

Algorithm 3 presents the steps of merging a join (split) association of open nodes of a synthetic graph and a split (join) association of open nodes of a component. If the associations are immediate, the pair of nodes are merged (Algorithm 2). If the associations are eventual, for each pair open nodes (one in

Algorithm 2. Merge a Pair of Open Nodes

input : a synthetic process graph: sg, a component: c, an open node of sg: n_{sg}, an open node of c: n_c, a synthetic label set: $\mathcal{SL}$
output: a synthetic process graphs: sg

1 **begin**
2 $N_{sg} \leftarrow N_{sg} \cup (N_c - \{n_c\})$;
3 **if** $n_{sg} \in preON_{sg} \wedge n_c \in postON_c$ **then**
4 $preON_{sg} \leftarrow (preON_{sg} - \{n_{sg}\}) \cup preON_c$;
5 $postON_{sg} \leftarrow postON_{sg} \cup (postON_c - \{n_c\})$;
6 **else if** $n_{sg} \in postON_{sg} \wedge n_c \in preON_c$ **then**
7 $preON_{sg} \leftarrow preON_{sg} \cup (preON_c - \{n_c\})$;
8 $postON_{sg} \leftarrow (postON_{sg} - \{n_{sg}\}) \cup postON_c$;
9 $E_{sg} \leftarrow E_{sg} \cup \{(n, n_3)|n_3 \in n_1\bullet\} \cup \{(n_3, n)|n_3 \in \bullet n_1\}$
10 $\cup(E_c - \{(n_1, n_3)|n_3 \in n_1\bullet\} - \{(n_3, n_1)|n_3 \in \bullet n_1\})$;
11 **foreach** $n \in (postON_c - \{n_c\})$ **do**
12 select a label sl from $\mathcal{SL}$ with probability $P(sl)$(Definition 16).
13 $\lambda_{sg} \leftarrow \lambda_{sg} \cup \{(n, sl)\}$;
14 **return** sg;

the synthetic graph and one in the component) a sequence component is selected to connect the pair of nodes. The connection is done by merging the pair of open nodes with two open nodes in the sequence component respectively (lines 7-11). The join (split) associations of open nodes are updated. The merged associations are deleted. If there are other split and join associations in the components, these associations are recorded in the synthetic graph (lines 12-15).

5 Evaluation

This section presents the evaluation of the algorithm in this paper. The algorithm is run 10 times with 604 SAP reference models as inputs. In total, 6040 synthetic models are generated in minutes. Table 1 shows the properties of the SAP reference models and the synthetic models. The properties includes average number of words in a label, average occurrences of a label, average number of nodes and edges in a process model, average number of local maximal split, join, loop, and sequence features in a model. From Table 1 we can know, the properties of the input collection (the SAP reference models) and the output collection (synthetic models) are very similar. The number of label occurrences is a bit larger in the synthetic models. This is because labels consisting of words with low frequencies may not be selected for any node in the synthetic models. The amount of labels are smaller, while the amount of nodes are almost the same. Therefore, label occurrences are higher in the synthetic models. The amount of splits are more in the synthetic models while the amount of joins are less. This is because during model generation post open nodes are considered first, which usually generates splits. The evaluation shows that the algorithms can indeed

Algorithm 3. Merge Join and Split Associations of Open Nodes

```
input  : a synthetic process graphs: sg, a component: c, a join (split)
         association of sg: AN_sg, a split (join) association of c: AN_c, a synthetic
         label set: SL
output: a synthetic process graphs: sg
1  begin
2      foreach n_sg ∈ AN_sg do
3          randomly select n_c from AN_c;
4          AN_c ← AN_c − {n_c};
5          if ζ_sg(AN_sg) = Im then
6              sg ←mergeNode (sg, c, n_sg, n_c, SL);//Algorithm 2.
7          else if ζ_sg(AN_sg) = Ev then
8              C_1 ← {c_1 ∈ C|Type(c_1) = sequence ∧ (∃(n_1 ∈ preON_c1 ∧ n_2 ∈
               postON_c1), n_1 → ... → n_2 is a subgraph of c_1)};
9              select c_1 ∈ C_1 with probability FCO(c_1, C_1) (Definition 14);
10             sg ←mergeNode (sg, c_1, n_sg, n_1, SL);//Algorithm 2.
11             sg ←mergeNode (sg, c, n_2, n_c, SL);//Algorithm 2.
12     Delete AN_sg from the domain of ζ_sg;
13     Delete AN_c from the domain of ζ_c;
14     foreach AN_1 ∈ {N_1 ⊆ N_c|ζ_c(N_1) ∈ {Im, Ev}} do
15         ζ_sg(AN_1) ← ζ_c(AN_1);
16     return sg;
```

be used to generate a much larger model collection that has similar properties as the model collection from which it is derived.

Table 1. Properties of Process Models

	Words	Label-Occurrences	Nodes	Edges	Splits	Joins	Sequences	Loops
SAP models	3.5	2.5	20.7	20.8	1.7	1.7	4.3	0.055
Synthetic models	3.1	3.2	20.3	22.7	2.0	1.4	4.3	0.055

6 Conclusion

This paper presents a technique that generates collections of synthetic process models based on a given process model collection. These collections can be used for the evaluation of different algorithms that are developed for large collections of business process models, such as algorithms for process similarity search, process querying, and version management. The evaluation shows that the algorithms can indeed be used to generate a much larger model collection that has similar properties as the model collection from which it is derived.

The techniques presented in this paper can be improved in several ways in future work. Firstly, while the technique is specifically developed to generate process model collections that are similar to a given collection in terms of some selected patterns. It can be improved to cover other patterns as well. Secondly, the technique in this paper mainly focuses on labels and control-flow. However, process models often contain more information, such as resources and data, which are also useful for detecting differences between models. The technique can be extended to cover other aspects as well.

References

1. Curran, T.A., Keller, G.: SAP R/3 Business Blueprint - Business Engineering mit den R/3-Referenzprozessen. Addison-Wesley, Bonn (1999)
2. Dijkman, R.M., Gfeller, B., Kster, J., Vlzer, H.: Identifying Refactoring Opportunities in Process Model Repositories. IST 53(9), 937–948 (2011)
3. Gao, X., Chen, Y., Ding, Z., et al.: Process Model Fragmentization, Clustering and Merging: An Empirical Study. In: BPM Workshops, Beijing, China (2013)
4. Kuramochi, M., Karypis, G.: Frequent Subgraph Discovery. In: ICDM (2001)
5. van Hee, K., La Rosa, M., Liu, Z., Sidorova, N.: Discovering characteristics of stochastic collections of process models. In: Rinderle-Ma, S., Toumani, F., Wolf, K. (eds.) BPM 2011. LNCS, vol. 6896, pp. 298–312. Springer, Heidelberg (2011)
6. ter Hofstede, A.H.M., Ouyang, C., La Rosa, M., Song, L., Wang, J., Polyvyanyy, A.: APQL: A process-model query language. In: Song, M., Wynn, M.T., Liu, J. (eds.) AP-BPM 2013. LNBIP, vol. 159, pp. 23–38. Springer, Heidelberg (2013)
7. La Rosa, M., Reijers, H.A., van der Aalst, W.M.P., Dijkman, R.M., Mendling, J., Dumas, M., Garcia-Banuelos, L.: APROMORE: An advanced process model repository. Expert Systems with Applications 38(6), 7029–7040 (2011)
8. Li, J., Wen, L., Wang, J., Yan, Z.: Process Model Storage Solutions: Proposition and Evaluation. In: Song, M., Wynn, M.T., Liu, J. (eds.) AP-BPM 2013. LNBIP, vol. 159, pp. 56–66. Springer, Heidelberg (2013)
9. Yan, Z., Dijkman, R.M., Grefen, P.W.P.J.: Business Process Model Repositories - Framework and Survey. Information and Software Technology 54(4), 380–395 (2012)
10. Yan, Z., Dijkman, R.M., Grefen, P.W.P.J.: Fast Business Process Similarity Search. Distributed and Parallal Databases 30(2), 105–144 (2012)
11. Yan, Z., Dijkman, R., Grefen, P.: FNet: An Index for Advanced Business Process Querying. In: Barros, A., Gal, A., Kindler, E. (eds.) BPM 2012. LNCS, vol. 7481, pp. 246–261. Springer, Heidelberg (2012)

A Method of Process Similarity Measure Based on Task Clustering Abstraction

Jian Chen, Yongjian Yan, Xingmei Liu, and Yang Yu*

School of Information Science and Technology, Sun Yat-Sen University, Guangzhou, Guangdong, P.R.China, 510006
yuy@mail.sysu.edu.cn

Abstract. A variety of methods have been proposed to measure the similarity of process models. But most of the methods only consider the structure while ignoring the semantic feature of the process model. When dealing with the process models with similar semantics but different in structure, these methods fail to achieve due similarity value. In this paper, a novel process model abstraction method is proposed which can keep the semantics as well as the structure features of the process model during abstraction. The output can then be used in the similarity measure. The experiment shows that this method can significantly improve the similarity value and make it closer to actual conditions.

Keywords: Process abstraction, process similarity, workflow management.

1 Introduction

Business Process Model is a precious intellectual property of many companies and has already become an indispensable part. A well-formed business process model can be widely used in office automation, e-government affairs and e-commerce etc. Large enterprise groups usually need to maintain thousands of business process models. Some of these models are abstracted from real business models and some are the reuse or reconstitution from the existed business models or even from mining the event logs.

To manage a large amount of business models, it is important to analyze the similarity and differentia among the models. This promotes the development of similarity match and retrieval. There have already existed many methods to analyze the similarity between business models. But many methods have the same problem that they fail to consider the semantics aspect when doing m:n node matching. This largely reduces the accuracy especially when dealing with business process models does not use uniform words or uniform business logical level to describe the same business process.

Actually, different people will build the same process model in different granularity. Two process models might have the same purpose but are not homogeneous. e.g. a process model A has a task "Ask for Leave" while process model B has a task

* Corresponding author.

C. Ouyang and J.-Y. Jung (Eds.): AP-BPM 2014, LNBIP 181, pp. 89–102, 2014.

sequence “Write a Vacation”, “Summit Application”, “Wait for Reply”. These two models have the same semantics but are described in different business logical level, and they fail to get due similarity value.

In this paper, a novel business process model abstraction method considering the business semantics is proposed. The abstraction method is based on clustering. Experiments show that the proposed method can obtain higher similarity value especially when dealing with those business model pairs which are actually similar but constructed in different logical level and fail to get a high similarity score. It proves that this method can make the similarity value closer to the actual conditions

The structure of this paper is as follows. Section 2 discusses some representative related works. Section 3 gives the definition of the similarity of process model. Section 4 introduces the process model abstraction method based on clustering. Section 5 examines the effectiveness of the method. Section 6 concludes this paper.

2 Related Work

The similarity matching methods can be divided into two categories, through the similarity of the process model and through similarity search. The evaluations of the similarity are mostly based on the similarity of the elements (nodes and edges), the graph edit distance and the behavior of the business process model.

[1] proposes a similarity measure method based on the similarity of the label of the nodes in process models. This method calculates the string edit distance [2] of each node’s label to decide the node mapping between two process models, then the similarity of two graphs are defined as the sum of all labels’ similarities. This method fails to consider the effect of node’s position. [3] proposes a semantics based similarity measure method. This method expresses the process model in semantics aspect. It first calculates the string edit distance of each node’s label, then uses WordNet[4], a well-known English dictionary, to obtain the synonyms and near-synonym between labels. Finally, it compares the attributes and relations between nodes. This method deals with the position of nodes only when the nodes’ labels are not equal or are not semantically similar. When two nodes with similar labels but are in different location of the process model, the similarity value is still high. [5] points out that comparing the similarity of each two process models in a process model database is very inefficient. It is more efficient to calculate the similarity by building a similar calculation index based on the feature of the process model. This method also first calculates the string edit distance between the nodes’ labels, then maps the nodes using different threshold value according to the roles of the nodes (roles are defined as the relationship of the nodes input and output edge), finally gets the similarity. [6] defines the similarity of process models as the ratio of the graph edit distance and the number of the elements in the model. The graph edit distance is the number of operations (insert, remove nodes and edges) in order to change a graph to another. [7] models the process model to a ordered tree structure, all the nodes are the leaves of the tree. Then using the tree edit distance to measure the similarity but it is unable to handle the loop structure. [8] defines the Minimum Common Supergraph and computes the similarity as the graph edit distance of Maximum Common Subgraph[9]. [10] further considers the nodes causal relationship and defines the behavioral profiles similarity. [11] builds

a vector space based on the causal relationship of the nodes and then measures the similarity in the space. [12] computes the similarity through the path set relationship of two process models. [13] separates the similarity into node similarity and dependence similarity. It models the nodes by a vector, and uses the vector distance as similarity. [14] uses the Complete Finite Prefix to represent a process model and compute the shortest succession distance between tasks matrix to get the similarity value.

3 Similarity Definition

The similarity measure has four steps. First, the tasks of the process model are clustered. Second, each of the clusters is analyzed in order to decide which tasks are able to be merged. Third, merge the tasks then reset the label and I/O conditions based on the results of the analysis. Finally, use any graph similarity measure algorithm to evaluate the effort.

First of all, the similarity definition of the process models is discussed in this section, and the abstraction method will be discussed in section 4.

All the workflow process models in this paper are modeled as Petri Net[15].

Definition 1 (Process Model). A workflow process model can be represented as a quintuple $\langle \mathbf{P}, \mathbf{T}, \mathbf{F}, \mathbf{i}, \mathbf{o} \rangle$ and:

- P is a finite set contains all the places.
- T is a finite set contains all the transitions, each transition represents a task in the workflow.
- i is the entrance of the process model and o is the exit of the process model.
- $F \subseteq (\{i\}) \times T) \cup (T \times \{o\}) \cup (T \times P) \cup (P \times T)$ represents the structure of the process model.

A workflow process model contains two types of information: (1) the dependency relation between tasks; (2) the correspondence among tasks, roles and resources;

The definition of roles and resources are simplified, as the roles and resources management is not the focus point of this paper. Notices that all the workflow process models mentioned in this paper have consistent control flow and data flow.

The target of this paper is to propose an abstract method so that can obtain higher similarity value of two process model after abstraction. The rest of this section will introduce the node similarity measuring method and the process model similarity measuring method.

3.1 Node Similarity

To measure the node similarity, the definition in [1] is used in this paper, it shows that the node similarity in the process model can be obtained from five aspects: syntax, semantics, attribute, type and contextual. The syntactic similarity considers the syntax aspect of the label to calculate the similarity. The semantic similarity considers the semantics of the words in label to calculate the similarity. The attribute similarity takes the task's attribute value and type into consideration to calculate the similarity between nodes. The type similarity calculates the nodes' similarity through the node's type and business characters. The contextual similarity combines the nodes and their execution sequence to calculate the similarity.

The syntactic similarity returns the string edit distance between the process model nodes, it is computed by counting the operation of insertion, deletion or replacement of charts in the string to make it become another string. Assumed that PLM_1, PLM_2 are two labels, $|PLM_1|$ is the length of the label, the function $Edit(|PLM_1|,|PLM_2|)$ is the string edit distance of PLM_1 and PLM_2, the syntactic similarity is counted as:

$$\text{syntacticSim}(PLM_1, PLM_2) = 1 - \frac{Edit(|PLM_1|,|PLM_2|)}{\max(|PLM_1|,|PLM_2|)} \tag{1}$$

To ensure the accuracy, all the letters are treated as lower case when computing the syntactic similarity.

The semantic similarity returns the similarity of the words in the label. Assumed that PLM_1 and PLM_2 are two labels, function $\text{Words}(PLM_1)$ partitions the labels into a word set, denoted as $WPLM_1 = \text{Words}(PLM_1)$, and $|WPLM_1|$ is the number of words in the word set. To improve the accuracy, the synonyms should be taken into consideration. Function $\text{synW}(WPLM_1, WPLM_2)$ is the intersection of the synonyms in $WPLM_1$ and $WPLM_2$, and wId is the weight of a word, wSy is the weight of a synonyms, so the semantic similarity can be defined as:

$$\text{semanticSim}(PLM_1, PLM_2) = \frac{2 * wId * |PLM_1 \cap PLM_2| + wSy * (|\text{synW}(WPLM_1, WPLM_2)| + |\text{synW}(WPLM_2, WPLM_1)|)}{|WPLM_1| + |WPLM_2|} \tag{2}$$

From the business aspect, a node in a process model may belong to different business type and affect the similarity value. Nodes belong to the same type will obtain higher similarity. Here, function $\text{type}(Node)$ is the type of a node and the type similarity function $\text{typeSim}(Node_1, Node_2)$ returns a boolean value which denotes whether the two nodes belong to the same type.

$$\text{typeSim} = \begin{cases} 0 & \text{type}(Node_1) \neq \text{type}(Node_2) \\ 1 & \text{type}(Node_1) = \text{type}(Node_2) \end{cases} \tag{3}$$

From the business semantics aspect, nodes belong to the same type may have different attributes. The attribute similarity first calculates the syntactic and semantic similarities of each pair of attributes, and then gets the weighted average of the results. Assumed that PLM_1^i is the ith attribute of PLM_1. PLM_1 and PLM_2 belong to the same type and both have n attributes, $i, j \in [1, n]$, the function attrSim(PLM_1, PLM_2) is defined as follow:

$$\mathrm{arrtriSim}(PLM_1, PLM_2) = \frac{\sum_{i=1}^{n} \mathrm{syntacticSim}(PLM_1^i, PLM_2^i) + \mathrm{semanticSim}(PLM_1^i, PLM_2^i)}{n} \quad (4)$$

In general, the attribute similarity will use together with the syntactic and semantic similarities to raise the accuracy of the nodes' similarity.

The contextual similarity takes the precursor and successor nodes into consideration when computing the nodes' similarity. Assumed that Pre_1 and Pre_2 are the set of precursor nodes of $Node_1$ and $Node_2$, $Succ_1$ and $Succ_2$ are the set of precursor nodes of $Node_1$ and $Node_2$. Function Msim($Nodes_1, Nodes_2$) is the best node mapping of the node set $Nodes_1$ and $Nodes_2$, the best node mapping is the mapping with the highest similarity value through the syntax, semantics, attribute and type similarity. The contextual similarity is defined as:

$$\mathrm{contextualSim}(Node_1, Node_2) = \frac{\mathrm{Msim}(Pre_1, Pre_2) * \sqrt{|Pre_2|}}{\sqrt{|Pre_1|}} + \frac{\mathrm{Msim}(Succ_1, Succ_2) * \sqrt{|Succ_2|}}{\sqrt{|Succ_1|}} \quad (5)$$

According to the five aspects of similarity introduced above, the similarity value of two nodes can be defined as:

$$nodeSim(n,m) = \sum_i w_i * sim_i(n,m), \sum w_i = 1$$
$$sim_i \in \{\mathrm{syntacticSim}, \mathrm{semanticSim}, \mathrm{typeSim}, \mathrm{attriSim}, \mathrm{contextualSim}\} \quad (6)$$

Here, w_i is the weight of each similarity.

3.2 Process Model Similarity

The process model similarity can be computed by using the node similarity. In this paper, the method in [6] is used to measure the similarity of two process models. It shows that the similarity of two process model $W_1 =< P_1, T_1, F_1, i_1, o_1 >$ and $W_2 =< P_2, T_2, F_2, i_2, o_2 >$ can be computed through the graph edit distance and generates a mapping $M : P_1 \rightarrow P_2, T_1 \rightarrow T_2$. So the structural similarity of W_1 and W_2 is:

$$structSim(W_1, W_2) = 1 - \frac{wskipn * skipp}{|p_1| + |p_2|} - \frac{wskipn * skipt}{|t_1| + |t_2|} - \frac{wskipe * skipe}{|f_1| + |f_2|} - \frac{wsub * 2 * \sum_{(n,m) \in M} 1 - nodeSim(n,m)}{|subNode|} \quad (7)$$

Here, *skipp* is the set of the skip places, *skipt* is the set of skip transitions and *skipt* is the set of skip edges, $\sum_{(n,m)\in M} 1 - nodeSim(n,m)$ is the similarity of skip nodes pairs (n,m) which satisfy the mapping M, $|subNode|$ is the total number of the skip nodes. $wskipn$, $wskipe$, $wsub$ is the weight for the skip nodes, edges and the number of the skip nodes, and $wskipn, wskipe, wsub \in [0,1]$.

The similarity of the process models can be computed by (7). The accuracy of the similarity value is largely influenced by the node mapping. In order to get a more reasonable node mapping, one useful way is to reduce the impact of the non-uniform semantics and business logical level. This can be done by abstracting the process model.

4 Abstraction Method

This section introduces the clustering based abstraction method. The clustering should be able to keep both semantic and structure feature of the process model. The method expects to cluster those nodes with the similar semantics and maintain the process's behavior as well.

4.1 Task Clustering

In the process model, each task should be represented by a vector so that any clustering algorithm based on vector can be used. The vector for each task should cover both the structure and semantic features.

The vector model proposed in this paper is able to handle three kinds of structures: sequence, branch and loop, denoted as $node_1 \rightarrow node_2$, $node_1 || node_2$, $node_1$<->$node_2$. The structure feature for a task is expressed by the number of tasks which have certain relation with it. Notice that a join or split somehow separates a workflow, so those tasks which are successors of a join/split will not count into the structure feature of a task which is precursor of the same join/split. For example, in fig.1, the structure feature for t3 and t5 is:

- t3.sequence = |{t7}| = 1
- t3.branch = |{t5, t6}| = 2
- t3.loop = |{t2, t4}| = 2
- t5.sequence = |{t6}| = 1
- t5.branch = |{t1, t2, t3, t4}| = 4
- t5.loop = |{ }| = 0

The semantic feature of a task is computed through the task's labels. The task vector can be defined through the structure and semantic feature.

Definition 2 (Task Vector). A task vector $\{v_0, v_1, v_2, \ldots, v_n\}$ is a n-dimension vector. v_0, v_1, v_2, is the sequence, branch and loop feature normalized TF-IDF value, the others is the TF-IDF values for words contained in the task's label.

K-means clustering algorithm is used in this paper, the initial k tasks are those tasks which are start tasks of a sequence structure. The reason is that in the general situation, tasks with similar semantics are always not so "far" from each other and usually appear in the same sequence.

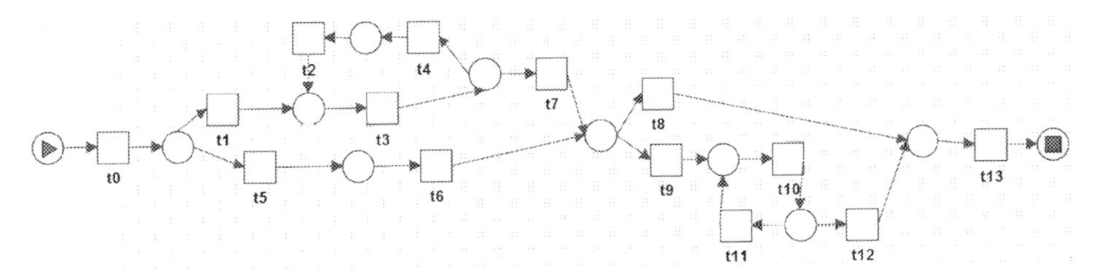

Fig. 1. A simple process model

4.2 Merge Rules

[16] gives a graph abstraction method using clustering algorithm. The method directly merges the nodes in every cluster which will clearly change the behavior and the structure of the process model, for not all the nodes in a cluster are able to be merged, e.g. two nodes may be very "far" from each other in the process model.

So in this paper, all clusters should be analyzed before merge. The analysis starts from the cluster which contains most tasks and then goes on to the other clusters. Four merge rules are introduced during the analysis:

Merge Rule 1 (Sequence Merge). If $node_1$ and $node_2$ in $Cluster_i$ and $node_1$->$node_2$, then merge $node_1$ and $node_2$

Merge Rule 2 (Branch Merge). If $node_1$ and $node_2$ in $Cluster_i$ and $node_1 \| node_2$, then merge $node_1$ and $node_2$.

Merge Rule 3 (Loop Merge). If $node_1$ and $node_2$ in $Cluster_i$ and $node_1$<->$node_2$, then merge $node_1$ and $node_2$.

Merge Rule 4 (Trace Merge). If $node_1$ and $node_2$ in $Cluster_i$ and there exists a task trace = [$node_1$,…,$node_2$] satisfies $|trace| <= threshold_1$.

Merge Rule 5 (Semantic Merge). If $node_1$ in $Cluster_i$, $node_2$ in $Cluster_j$ and semantic ($Cluster_i$, $Cluster_j$) <= $threshold_2$.

After analysis, the nodes in the same cluster which satisfy one of the five rules will be merged into a new node. The new node represents a task in the abstracted process model. The label of this task is the union set of the merge nodes. The input condition is set the same as the "front" most nodes (with no other nodes in the merged nodes are precursors of it) while the output condition is set the same as the "behind" most nodes (with no other nodes in the merged nodes are successor of it). e.g. fig.3 is the abstraction graph of fig.2

Finally, after merging, the abstract process model is generated and can be used to compute the similarity value by (7).

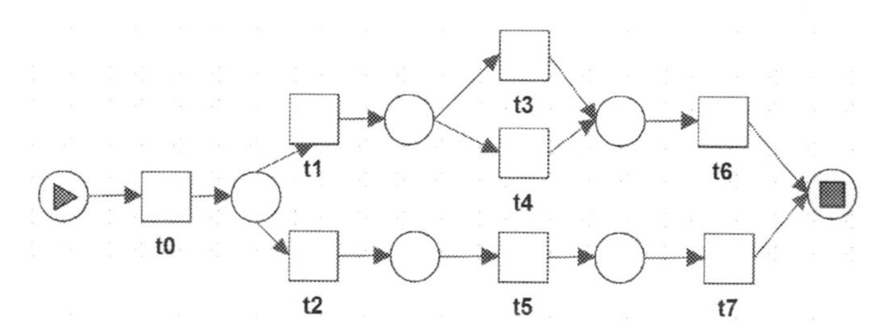

Fig. 2. Original process Model

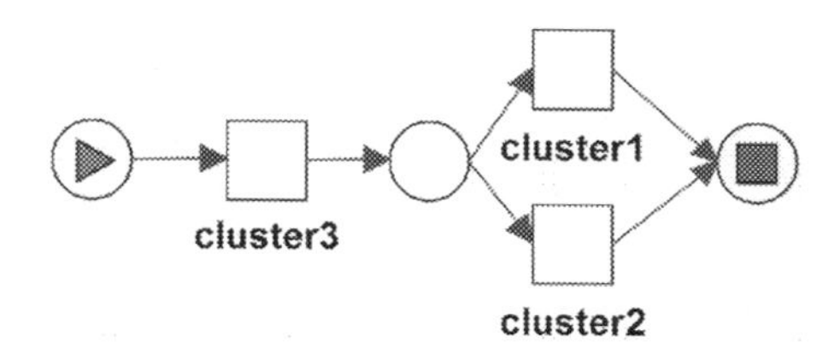

Fig. 3. Abstracted process Model, *cluser1*={*t1*,*t3*,*t4*,*t6*}, *cluster2*={*t2*,*t5*,*t7*}, *cluster3*={*t0*}

4.3 An Example

In order to better illustrate the abstraction method, an example process model is given here in fig.4. This example will show the abstraction method step by step.

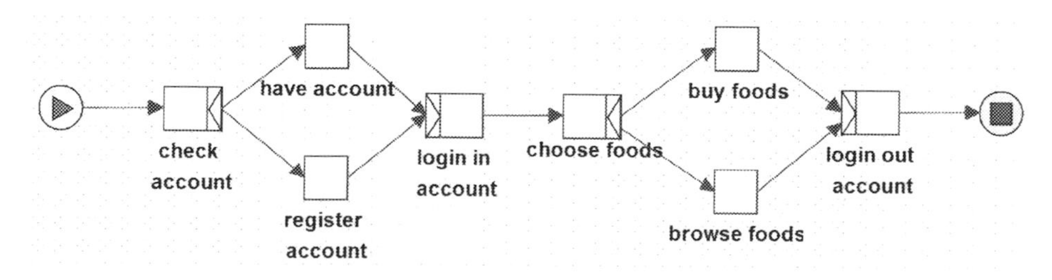

Fig. 4. Original process model

Given a process model, the words in all the tasks' label are processed first. All the works are reduce to their stem by stemming algorithm [17], e.g. choose → choos. Then the synonym table is made based on WordNet. The synonym table records all the synonym relationship among words, which can make the TFIDF more accuracy in the next step.

The next step is to do the task clustering using Kmeans. Firstly the tfidf value for each task needs to be calculated. The tfidf is consisted of two parts, the tfidf value for the words and for the structure. The tf value for words in a task is calculated by $word.app / total$, here word.app is the times that a word appears in the task's label and total is the number of words in the label. The idf value for words is calculated by $\log(all / contains)$, here *all* is the total number of tasks and *contains* is the number of tasks which contains the word. For example, to calculate the tfidf value for the word "account" in task *have account* is $(1/2)*\log(8/5) = 0.102$. The tfidf value for structure is calculated in similar way.

Table 1. Clustering result

Cluster Id	Task (represented by their label)
1	"check account" ,"have account", "register account", "login in account", "login out account"
2	"choose foods", "buy foods", "browse foods"

After calculating the tfidf value for every word, a vector can be built for each task. Based on the vectors, the Kmeans algorithm can be used here to get the task clusters. The clustering result is shown in table 1.

Here the structure is simple so it does not have an obvious impact on the clustering result.

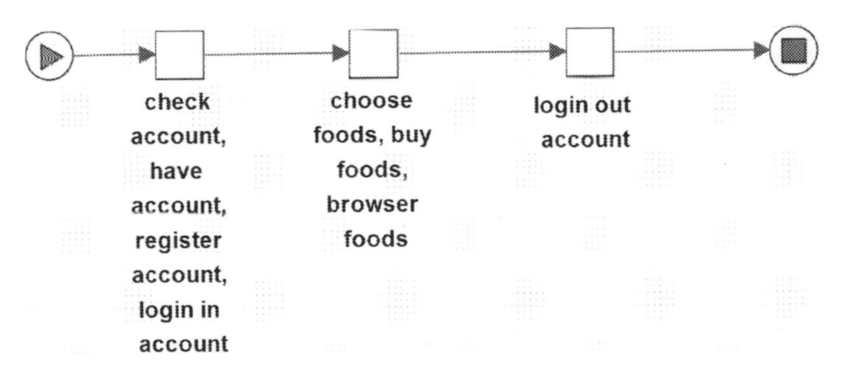

Fig. 5. The final process model after clustering and merge analyzing

Finally, each cluster is analyzed with the merging rules, those tasks in the same cluster and satisfy the merging rules are merged into a new task. The label of the new task is the union set of the merged tasks' labels. Here in the first task, the task *check account* is the "front" most task and the *login in account* is the "behind" most task. So the I/O conditions are set follow this two tasks. The final result is shown in fig.5. Notices that the task *login out account* fails to meet the merge rules so it is not included in the new merged task.

5 Experiments

Comparison experiments are conducted in this section to explain the improvement of process model similarity by using abstraction method proposed in section 4. The dataset used in the experiments are from the students activities of the summer campus in SYSU, the students are asked to build a process model with the similar purpose. The largest process model in the dataset has 55 tasks while the smallest has 12 tasks. Most process models have about 20 to 30 tasks.

First the experiments are conducted in order to prove that the abstraction method can make the similarity value higher when dealing with those process model pairs have similar purpose but are not homogeneous. 14 pairs out of the 56 process models are chosen and validated by 3 experts that all pairs ought to have high similarity value.

The similarity algorithm using here is based on the algorithm proposed in [6], which has been a part of proM[1]. The algorithm is modified according to the definition in section 3. There are three similarity measure algorithms, the greedy algorithm, the exhaustive algorithm and the A* algorithm.

5.1 Using Greedy Similarity Algorithm

The similarity value using greedy similarity algorithm is shown in fig.6, the result shows that in all the 14 cases, higher similarity value can be obtained after using the clustering abstraction method. Especially in case 2, 6 and 11, there are significant improvements. This is mainly because the time complexity of the greedy algorithm is $O(N^3)$, it always choose the best node mapping to achieve the shortest graph edit distance in every step. In many cases, the obtained solution is a second-best-solution. When the task is merged through the abstraction method in section 4, the process

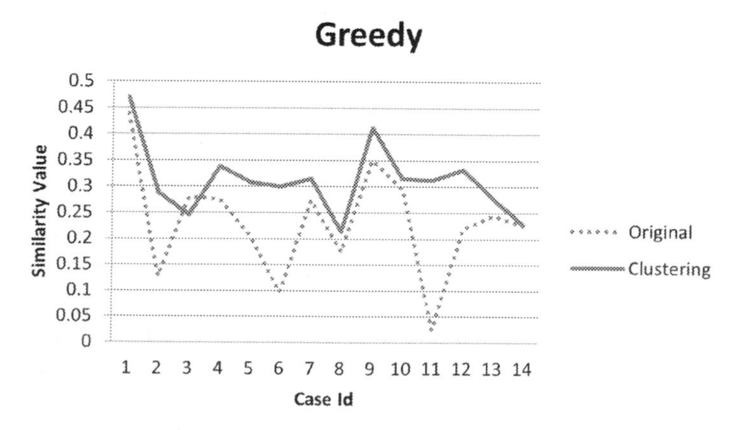

Fig. 6. Similarity value using greedy algorithm

[1] http://www.promtools.org/prom6/

model can better reflect the structure of the business process and make the second-best-solution better.

5.2 Using Exhaustive Algorithm with Pruning

The similarity value through the exhaustive similarity algorithm is shown in fig.7. The result is not as well as the greedy one, but still has some improvement. This is because the exhaustive algorithm first takes all the node pairs which satisfy the semantic similarity to a mapping set, and finally keeps some pairs of nodes that achieve high semantic similarity value. After clustering and abstraction, the exhaustive algorithm can better reflect the business process semantics and only consider the node pair with high semantic similarity value. So the algorithm fails to get a high discrimination when dealing with the process models with similar structure.

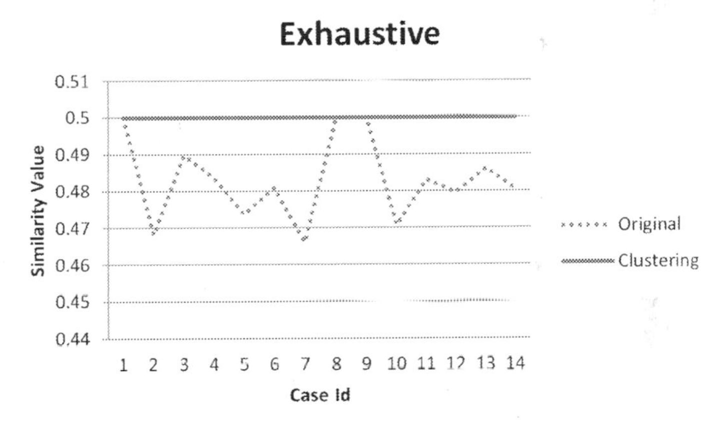

Fig. 7. Similarity value using exhaustive algorithm

5.3 Using A* Similarity Algorithm

The similarity value using the A* similarity algorithm is shown in fig.8, it has significant improvement after clustering and merging the tasks. This is mainly because in every iteration of the A* algorithm, the evaluation function will compare the similarity of the chosen node pairs. Those pairs with higher similarity value will be handled first and the A* algorithm limits the number of the node pair as well. If the original process model have great difference in semantics and structure between nodes in the initial stage, after the clustering and abstraction, both the structure and semantics of the models are more similar. So the improvement of this algorithm in similarity value is more obvious

These three experiments show that after using the clustering and abstraction method proposed in this paper, higher similarity value can be obtained when dealing with those process models consider to be similar. The fact is that, in large enterprise, many business processes have similar semantics so using this method can discover

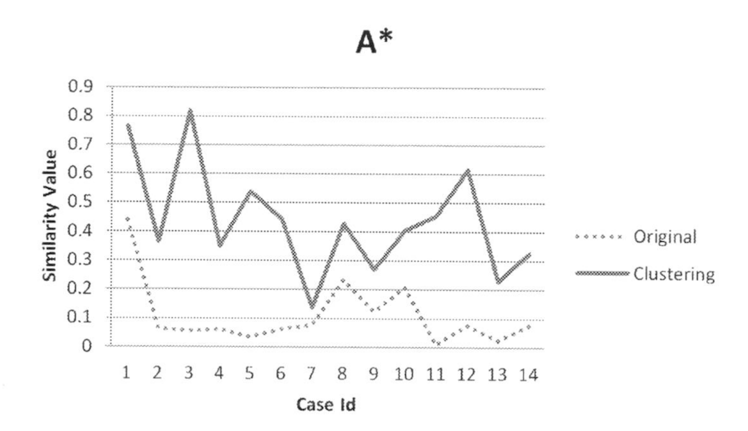

Fig. 8. Similarity value using A* algorithm

more similar business processes. The result can be further used in structure reused, frequent substructure mining etc.

5.4 Precision and Recall Experiments

The next experiment is done in a process model set and using the precision and recall to evaluate the effect.

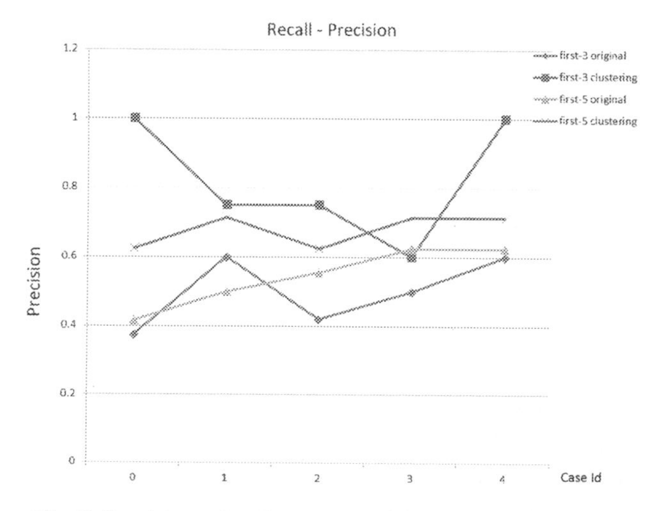

Fig. 9. Precision using 5 process models as example queries

The process model set contains 56 process models (280 pairs). The similarity value is evaluate by 3 workflow experts and ranked on a 1 to 7 Linker scale. Those pairs ranked 5 or more are considered to be similar.

5 process models are using as example queries to compare with the models in the dataset; the returned similarity result is then sorted to compute the precision. The greedy algorithm is used here and the experiment evaluates the precision when selecting the first 3 or 5 best models.

As in fig.9 shows, the similarity is improved after using the clustering method in all cases. This experiment proves that the clustering abstraction method not only raises the similarity but makes the similarity value closer to the actual condition.

6 Conclusion

In this paper, we propose a process model abstraction method which can remain the behavior and structure information of the original model. The novelty of the method includes:

1. The task vector can represent the structure feature of the process model. These features include sequence, loop and branch and can affect the clustering to get a more reasonable result.
2. Five merge rules are proposed here to analyze the cluster result. These rules can maintain the behavior of the process model.

Through experiments, this method can discover those process models respected to be similar but fail to obtain a high similarity value because of the huge difference in structure. The output can be further used in many areas of business process management. There are still some drawbacks in the abstraction method:

1. The task vector just invokes the structure of the model in a simple way, and it may not reflect the structure of the context of the task clearly. So further modeling work is needed to better model the structure.
2. The data set in this paper is with limited size, when facing huge datasets, the efficiency of the clustering may be a bottle neck in practices.

Acknowledgments. This work is Supported by the National Natural Science Foundation of China under Grant No.60873162; the Natural Science Foundation of Guangdong Province under Grant No.S2012010009634; the Research Foundation of Science and Technology Major Project in Guangdong Province under Grant No.2012A010800012.the Cooperation Project in Industry, Education and Research of Guangdong Province & Ministry of Education of China under Grant No.2011B090400275; the Research Foundation of Science and Technology Plan Project in Guangzhou City under Grant No.12A12051586.

References

1. Dijkman, R., Dumas, M., van Dongen, B., et al.: Similarity of business process models: Metrics and evaluation. Information Systems 36(2SI), 498–516 (2011)
2. Levenshtein, V.I.: Binary Codes Capable of Correcting Deletions, Insertions and Reversals (1966)

3. Ehrig, M., Koschmider, A., Oberweis, A.: Measuring similarity between semantic business process models. In: Proceedings of the Fourth Asia-Pacific Conference on Comceptual Modelling, vol. 67, pp. 71–80. Australian Computer Society, Inc. (2007)
4. Miller, G.A.: WordNet: A lexical database for English. Communications of the ACM 38(11), 39–41 (1995)
5. Yan, Z., Dijkman, R., Grefen, P.: Fast Business Process Similarity Search with Feature-Based Similarity Estimation. In: Meersman, R., Dillon, T.S., Herrero, P. (eds.) OTM 2010. LNCS, vol. 6426, pp. 60–77. Springer, Heidelberg (2010)
6. Dijkman, R., Dumas, M., García-Bañuelos, L.: Graph Matching Algorithms for Business Process Model Similarity Search. In: Dayal, U., Eder, J., Koehler, J., Reijers, H.A. (eds.) BPM 2009. LNCS, vol. 5701, pp. 48–63. Springer, Heidelberg (2009)
7. Bae, J., Caverlee, J., Liu, L., Yan, H.: Process Mining by Measuring Process Block Similarity. In: Eder, J., Dustdar, S. (eds.) BPM Workshops 2006. LNCS, vol. 4103, pp. 141–152. Springer, Heidelberg (2006)
8. Fernández, M.-L., Valiente, G.: A graph distance metric combining maximum common subgraph and minimum common supergraph. Pattern Recognition Letters 22(6-7), 753–758 (2001)
9. Bunke, H., Shearer, K.: A graph distance metric based on the maximal common subgraph. Pattern Recognition Letters 19(3-4), 255–259 (1998)
10. Kunze, M., Weidlich, M., Weske, M.: Behavioral similarity: A proper metric. In: Rinderle-Ma, S., Toumani, F., Wolf, K. (eds.) BPM 2011. LNCS, vol. 6896, pp. 166–181. Springer, Heidelberg (2011)
11. van Dongen, B.F., Dijkman, R., Mendling, J.: Measuring Similarity between Business Process Models. In: Bellahsène, Z., Léonard, M. (eds.) CAiSE 2008. LNCS, vol. 5074, pp. 450–464. Springer, Heidelberg (2008)
12. Gerke, K., Cardoso, J., Claus, A.: Measuring the Compliance of Processes with Reference Models. In: Meersman, R., Dillon, T., Herrero, P. (eds.) OTM 2009, Part I. LNCS, vol. 5870, pp. 76–93. Springer, Heidelberg (2009)
13. Jung, J.-Y., Bae, J.: Workflow Clustering Method Based on Process Similarity. In: Gavrilova, M.L., Gervasi, O., Kumar, V., Tan, C.J.K., Taniar, D., Laganá, A., Mun, Y., Choo, H. (eds.) ICCSA 2006. LNCS, vol. 3981, pp. 379–389. Springer, Heidelberg (2006)
14. Wang, S., Wen, L., Wei, D., Wang, J.: SSDT matrix-based behavioral similarity algorithm for process models. Computer Integrated Manufacturing Systems 19(8), 1822–1831 (2013)
15. Yuan, C.: The Principle and Application of Petri Net. Publishing House of Electronics Industry (2005) (in Chinese)
16. Smirnov, S., Reijers, H.A., Weske, M.: A semantic approach for business process model abstraction. In: Mouratidis, H., Rolland, C. (eds.) CAiSE 2011. LNCS, vol. 6741, pp. 497–511. Springer, Heidelberg (2011)
17. Porter, M.F.: An algorithm for suffix stripping. Program: Electronic library and information systems 14(3), 130–137 (1980)

Efficient Syntactic Process Difference Detection Using Flexible Feature Matching

Keqiang Liu[1], Zhiqiang Yan[1,2], Yuquan Wang[2], Lijie Wen[2], and Jianmin Wang[2]

[1] Capital University of Economics and Business, Beijing, China
[2] Tsinghua University, Beijing, China
zhiqiang.yan.1983@gmail.com

Abstract. Nowadays, business process management plays an important role in the management of organizations. More and more organizations describe their operations as business processes It is common for organizations to have collections of thousands of business process models. The same process is usually modeled differently due to the different rules or habits of different organizations and departments. Even in the subsidiaries of the same organization, process models vary from each other, because these process models are redesigned from time to time to continuously enhance the efficiency of management and operations. Therefore, techniques are required to analyze differences between similar process models. Current techniques can detect operations required to modify one process model to the other. However, these operations are based on activities and the syntactic meanings are limited. In this paper, we define differences based on workflow patterns and propose a technique to detect these differences efficiently. The experiment shows that these differences indeed exist in real-life process models and are useful to analyze differences between business process models.

1 Introduction

Nowadays, organizations tend to enhance their management efficiency with the technology of business process management (BPM). More and more business processes are described as process models to facilitate the implementation of BPM. Therefore, it is common to see thousands of process models in an organization or even in a department. For example, the information department of China Mobile Communication Corporation (CMCC) maintains more than 8,000 processes in its BPM systems [10]. To manage such a large number of process models efficiently and automatically, business process model repositories are required [16,22,23]. These repositories provide techniques include detecting differences between process models [4], process similarity search [24] and process querying [25].

This paper focuses on detecting (syntactic) differences between business process models. Difference detection can, for example in case of merger between BPM systems of two or more organizations, be applied to detecting differences

C. Ouyang and J.-Y. Jung (Eds.): AP-BPM 2014, LNBIP 181, pp. 103–116, 2014.

between two process models describing the same operation in different organizations. For example, originally each of the 34 subsidiaries of CMCC built its own BPM systems to maintain its business process, but later the headquarters decided to build a unified one to maintain business process models of all subsidiaries [10]. Since these subsidiaries have almost the same business processes, these process models are similar with some variations. For each business process, the headquarters also would like to have one model that works for different scenarios of all subsidiaries instead of to maintain different models for different scenarios. Therefore, it is required to have a technique that can detect the differences between these similar process models. A further application of difference detection is process model editing. In the case of modifying one process model to anther, we can edit one fragment related to a difference instead of operating one activity at a time.

Figure 1 shows two similar business process models describing a "receiving goods" process. Both models are in the BPMN notation. Given two similar process models, the technique in this paper identifies the difference patterns between these models. For example, in Figure 1, activity *"Inbound Delivery Entered"* in *Model a* and activity *"Inbound Delivery Created"* in *Model b* are interchanged activities; F1.1 of *Model a* has an additional dependency, *"Invoice received"*, compared with F2.1 of *Model b*; F1.2 of *Model a* has different conditions compared with F2.2 of *Model b* because of different gateways. The technique in this paper detects these differences between process models efficiently.

There currently exists techniques [17,15] that can analyze syntactic differences between process models. However, the focus of these techniques are computing the operations of modifying one process model to another. In this paper, the

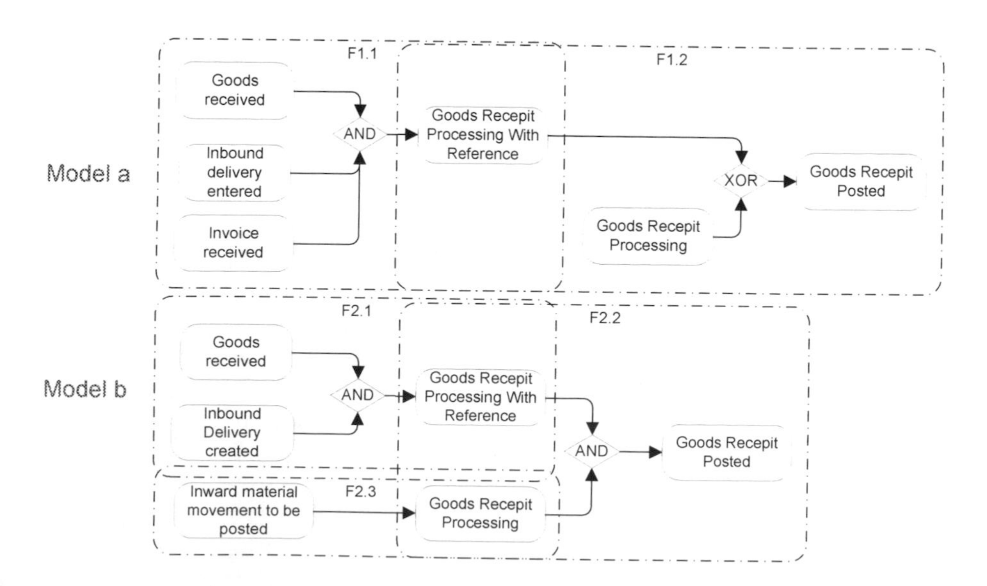

Fig. 1. Two business process models about receiving goods in BPMN

technique focuses on differences based on common workflow patterns, therefore provides more syntactic meanings about differences between process models. The experiment shows that the differences presented in this paper exists in real-life process models. Therefore, they are valuable for analyzing business process models.

The contribution of this paper is as following:

- syntactic differences are defined between business process models;
- flexible feature matching is proposed as a basis of difference detection;
- an algorithm is presented to show the steps to detect differences using flexible feature matching.

The rest of the paper is organized as follows. Section 2 presents preliminaries of the technique of this paper. Section 3 and 4 presents exact and flexible feature matching respectively. Section 5 presents different types of syntactic differences and shows how to detect these differences using flexible feature matching. Section 6 presents the steps of detecting differences between two business process models. Section 7 presents the experiment. Section 8 presents related work and section 9 concludes the paper.

2 Preliminaries

The technique in this paper is presented in the context of business process. A business process graph is a graph representation of a business process model [7,16], as described in Definition 1. As such, it focuses purely on the structure of that model, while abstracting from other aspects, e.g., different types of process modeling notations (BPMN, EPC, Petri net, etc.). Our detection of differences is based on the labels and structure of a business process model, therefore, abstracting from these aspects is acceptable. Due to the abstraction, types of nodes are also ignored, but we can still know which node is a gateway node from its label, e.g., xor, and from the size of its pre-set or post-set, e.g., a node has multiple nodes in its post-set. For example, Figure 2 shows the business process graphs for the models from Figure 1.

Definition 1 (Business Process Graph, Pre-set, Post-set). *Let $\mathcal{L}$ be a set of labels. A business process graph is a tuple (N, E, λ), in which:*

- *$N = (A, GW)$ is the set of nodes, which consists of a set of activity nodes A and a set of gateway nodes GW;*
- *$E \subseteq N \times N$ is the set of edges; and*
- *$\lambda : N \rightarrow \mathcal{L}$ is a function that maps nodes to labels.*

Let $G = (N, E, \lambda)$ be a business process graph and $n \in N$ be a node: $\bullet n = \{m|(m, n) \in E\}$ is the pre-set of n, while $n\bullet = \{m|(n, m) \in E\}$ is the post-set of n.

We consider features based on the most common workflow patterns: sequence, split, join, and loop. Definition 2 provides a formal description of these features [23].

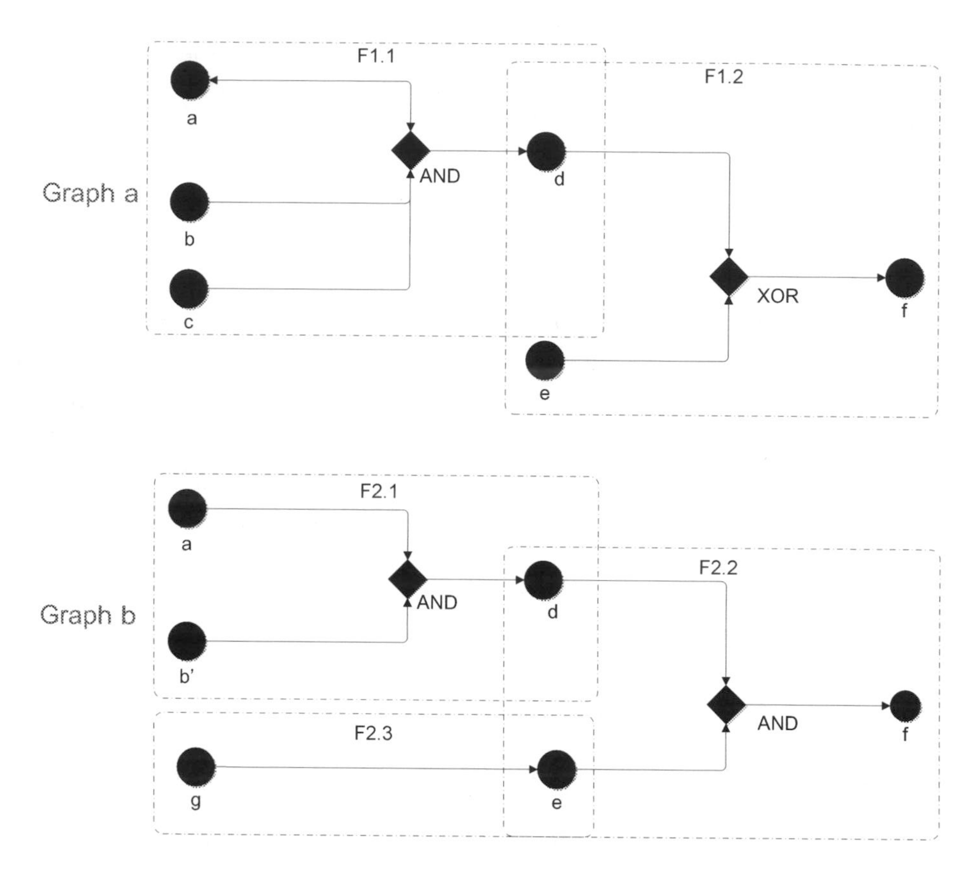

Fig. 2. Business process graphs

Definition 2 (Feature). *Let $g = (N, E, \lambda)$ be a process graph. A feature f of G is a subgraph of g. The size of a feature is the number of edges it contains. A feature is a:*

- *sequence feature of size $s - 1$ consisting of nodes $\{n_1, n_2, n_3, \ldots, n_s\}$, if E_f is the minimal set containing $(n_1, n_2), (n_2, n_3), \ldots, (n_{s-1}, n_s)$, for $s \geq 2$;*
- *split feature of size $s + 1$ consisting of a split node n, a node m and a set of nodes $\{n_1, n_2, \ldots, n_s\}$, if and only if E_f is the minimal set containing $(m, n), (n, n_1), (n, n_2), \ldots, (n, n_s)$, for $s \geq 2$;*
- *join feature of size $s + 1$ consisting of a join node n, a node m and a set of nodes $\{n_1, n_2, \ldots, n_s\}$, if and only if E_f is the minimal set containing $(n_1, n), (n_2, n), \ldots, (n_s, n), (n, m)$, for $s \geq 2$;*
- *loop feature of size s consisting of nodes $\{n_1, n_2, \ldots, n_s\}$, if E_f is the minimal set containing $(n_1, n_2), \ldots, (n_{s-1}, n_s), and (n_s, n_1)$, for $s \geq 1$.*

Some feature can further be decomposed into features, e.g., *feature F1.1* contains three subgraphs that are join features with two branches. However, we do not need to reconsider these subgraphs when we detect differences. Therefore, only local maximal features, as defined in Definition 3 need to be gotten from process graphs. For example, in Figure 2 features of *Graph a* and *Graph b* are circled in dash-dotted boxes. There are one local maximal sequence feature (*F2.3*) and four local maximal join features.

Definition 3 (Local Maximal Feature). *Let $g = ((A, GW), E, \lambda)$ be a process graph. A subgraph of g $f = ((A_1, GW_1), E_1, \lambda)$ is a feature. Feature f is a local maximal feature if and only if there exists no subgraph f' in g, such that f' is a feature of g and f is a subgraph of f'.*

We also notice that multiple split/join gateways may connect to each other as shown in Figure 3. Differences may occur between fragments like that. For example, the two fragments in Figure 3 have different conditions because they consist of the same set of activities but have different behaviors. Therefore, we define complex split and join features to be able to identify differences in complex structures like that.

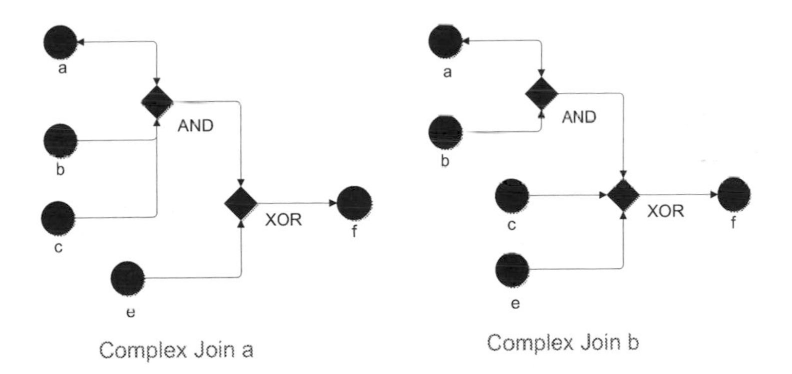

Fig. 3. Complex Join Features

Definition 4 (Complex Split/Join Feature). *Let $g = ((A, GW), E, \lambda)$ be a process graph. A subgraph of g $f = ((A_1, GW_1), E_1, \lambda)$ is a complex split feature if and only if:*

- $GW_1 \subseteq GW$ $(|GW_1| > 1)$, $\forall gw \in GW_1, |\bullet gw| = 1 \wedge |gw \bullet| \geq 2 \wedge (\exists gw_1 \in GW_1 \wedge gw_1 \neq gw$, *such that* $gw_1 \in gw \bullet \vee gw_1 \in \bullet gw)$;
- $A_1 \subseteq A$ $(|A_1| > 2)$, $\forall a \in A_1, \exists gw \in GW_1$, *such that* $a \in gw \bullet \vee a \in \bullet gw$;
- $E_1 = E \cap ((A_1 \cup GW_1) \times (A_1 \cup GW_1))$.

Complex join feature can be defined analogously.

3 Exact Feature Matching

We analyze nodes based on their label similarities. Label similarity can be measured in a number of different ways [9]. For illustration purposes we will use a syntactic similarity metric, which is based on string edit-distance, in this paper. However, more advanced metrics can be used that take synonyms and stemming [9] and, if possible, domain ontologies into account [13]. The label similarity and label matching are defined in the former work [5,24].

Definition 5 (Label Similarity, Node Matching). *Let $G = (N, E, \lambda)$ be a business process graph and $n, m \in N$ be two nodes and let $|x|$ represent the number of characters in a label x. The string edit distance of the labels $\lambda(n)$ and $\lambda(m)$ of the nodes, denoted $\mathrm{ed}(\lambda(n), \lambda(m))$ is the minimal number of atomic string operations needed to transform $\lambda(n)$ into $\lambda(m)$ or vice versa. The atomic string operations are: inserting a character, deleting a character or substituting a character for another. The label feature similarity of $\lambda(n)$ and $\lambda(m)$, denoted $\mathrm{lsim}(n, m)$ is:*

$$\mathrm{lsim}(n, m) = 1.0 - \frac{\mathrm{ed}(\lambda(n), \lambda(m))}{\max(|\lambda(n)|, |\lambda(m)|)}$$

Let lcutoff *be a cutoff value for node matching, which can be assigned as desired. Given two nodes n and m, they are matched, denoted as* $\mathrm{match}(n_1, n_2)$*, if and only if their label similarity is no less than* lcutoff*, i.e.,* $\mathrm{lsim}(n, m) \geq \mathrm{lcutoff}$.

For example, the similarity of label *"Inbound Delivery Entered"* and label *"Inbound Delivery Created"* is $1.0 - \frac{5}{24} = 0.79$. They are matched given a cutoff 0.75.

We consider two structural features to be matched, if their components (nodes and edges) are matched [24].

Definition 6 (Exact Feature Matching). *Given two features $f_1 = (N_1, E_1, \lambda_1)$ and $f_2 = (N_2, E_2, \lambda_2)$, they are matched, denoted as* $\mathrm{match}(f_1, f_2)$ *if and only if, there exists a bijection* $\mathrm{M} : N_1 \rightarrow N_2$*, such that*

- $|N_1| = |N_2|$;
- $\forall n \in N_1,\ \mathrm{match}(n, M(n))$;
- $|E_1| = |E_2|$;
- $\forall (n_1, n_2) \in E_1,\ (M(n_1), M(n_2)) \in E_2$.

For example, two sequence features of size s with lists of nodes $Ln = [n_1, n_2, n_3, \ldots, n_s]$ and $Lm = [m_1, m_2, m_3, \ldots, m_s]$ are matched if and only if for each $1 \leq i \leq s$: the node features of n_i and m_i are matched.

4 Flexible Feature Matching

In the previous section, we explain features and their exact matching. We know that the rules for exact feature matching are strict in their structure, i.e., two features of different types or sizes are never matched with each other. This can help identify the common features in two process graphs but makes it difficult to identify the differences. Therefore, in this section, we define two types of flexible feature matching and in the next section we show how to solve this issue with flexible feature matchings.

The first type is called the subgraph-flexible feature matching. In this type, two features are not matched, but there exists a subgraph of one feature is matched with a subgraph of the other. For example, in Figure 2 *F1.1* of *Graph a* and *F2.1* of *Graph a* are subgraph-flexible feature matching.

Definition 7 (Subgraph-flexible Feature Matching). *Given two features* $f_1 = ((A_1, GW_1), E_1, \lambda)$ *and* $f_2 = ((A_2, GW_2), E_2, \lambda)$*, they are subgraph-flexible matched, denoted as* $\text{Fmatch}_{\text{Sub}}(f_1, f_2)$ *if and only if, there exists a subgraph* $f'_1 = ((A'_1, GW'_1), E'_1, \lambda)$ *of* f_1 *and a subgraph* $f'_2 = ((A'_2, GW'_2), E'_2, \lambda)$ *of* f_2*, such that*

- $|A'_1| = |A'_2| > 0$;
- $\text{match}(f'_1, f'_2)$;
- $(\exists n_1 \in (A_1 - A'_1), \nexists n_2 \in (A_2 - A'_2), \text{match}(\text{n}_1, \text{n}_2)) \vee (\exists n_2 \in (A_2 - A'_2), \nexists n_1 \in (N_1 - N'_1), \text{match}(\text{n}_1, \text{n}_2))$.

The second type is the gateway-flexible feature matching. In this type, there exists an one-to-one mapping between activities of two features and each pair of mapped activities are matched. For example, in Figure 2, *F1.2* of *Graph a* and *F2.2* of *Graph b* are gateway-flexible feature matching.

Definition 8 (Gateway-flexible Feature Matching). *Given two features* $f_1 = ((A_1, GW_1), E_1, \lambda)$ *and* $f_2 = ((A_2, GW_2), E_2, \lambda)$*,* f_1 *and* f_2 *are gateway-flexible matching, denoted as* $\text{Fmatch}_{\text{Gate}}(f_1, f_2)$*, if and only if there exists a bijection* $\text{M} : A'_1 \rightarrow A'_2$*, such that*

- $|A'_1| = |A'_2|$;
- $\forall a \in A'_1,\ \text{match}(a, M(a))$;

5 Detecting Differences with Flexible Feature Matching

This section presents different types of syntactic differences between process models considered in this paper and how to detect these differences using flexible feature matching.

Firstly, we present differences on nodes, which are basis of other differences. Given two business process graphs, assuming that an activity does not have identical label with activities in the other graph. The activity is a skipped activity if no activities are matched with it, otherwise it is an interchanged activity. For example, in Figure 2, activity c (*"Invoice received"*) is a skipped activity; activity *b* (*"Inbound Delivery Entered"*) and activity *b'* *"Inbound Delivery Created"* are interchanged activities.

Definition 9 (Skipped Activity, Interchanged Activities). *Let* $g_1 = (N_1, E_1, \lambda)$ *and* $g_2 = (N_2, E_2, \lambda)$ *be two business process graphs. Given a node* $n_1 \in N_1$, n_1 *is a skipped activity, if and only if there exists no node* $n_2 \in N_2$ *matches with* n_1, *i.e.,* $\nexists n_2 \in N_2, \text{match}(n_1, n_2)$; n_1 *and* n_2 *are interchanged activities, if and only if there exists a node* $n_2 \in N_2$ *labeled differently matches with* n_1, *i.e.,* $\exists n_2 \in N_2, \text{match}(n_1, n_2) \wedge \lambda(n_1) \neq \lambda(n_2)$.

Then, we present differences on features, including additional branches, different branches, different order, different routing, and loop2once-off. Figure 4 shows an example for each type of difference considered in this paper. In the following content, each difference and how to detect it with features are explained.

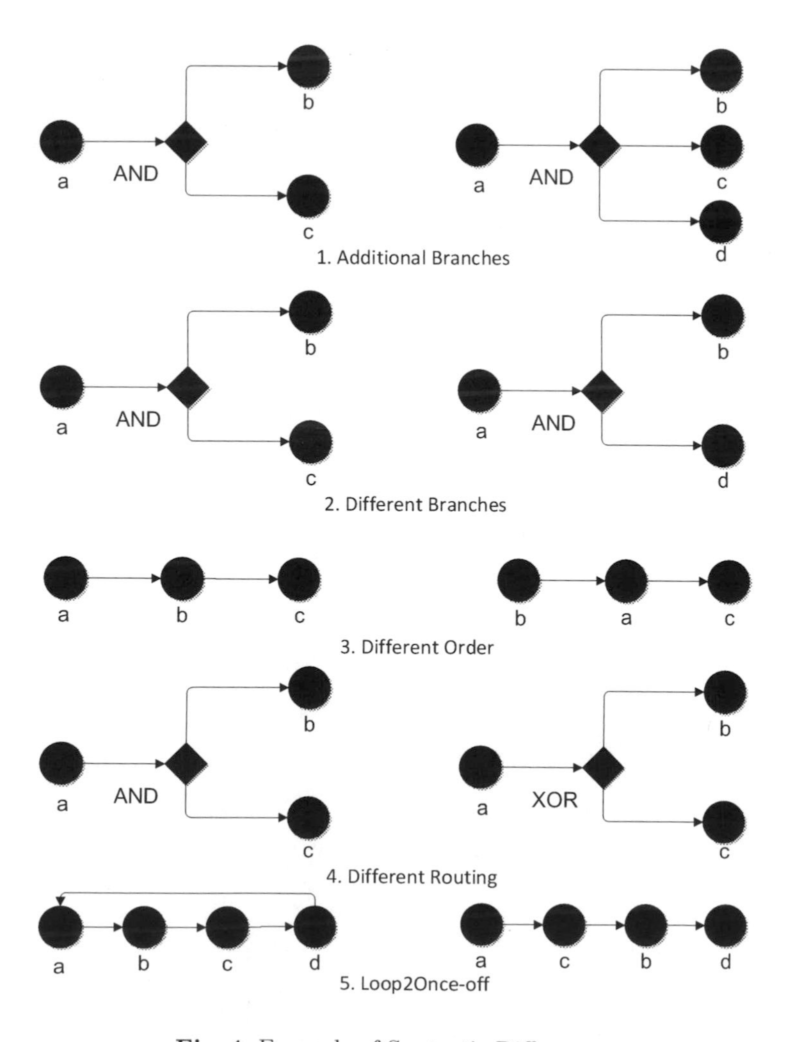

Fig. 4. Example of Syntactic Differences

5.1 Differences Related to Subgraph-Flexible Feature Matching

This sub-section describes two types of differences that can be detected using subgraph-flexible feature matching. Difference *additional branches* indicates that one join/split feature has additional branch(es) than the other join/split or sequence feature. For example, the first pair of features in Figure 4 is detected as *additional branches*, because the feature on the right has one more branch *d*. Difference *different branches* indicates that one join/split feature has different branch(es) with the other join/split feature. For example, the second pair of features in Figure 4 is detected as *different branches*, because branch *c* is only in the feature on the left while branch *d* is only in the feature on the right. Note that difference *additional branches* is a special case of *different branches*. We do not consider differences repeatedly between features. That is if two features are detected as *additional branches* then they are not detected as *different branches* anymore.

Definition 10 (Additional Branches, Different Branches). *Given two features f_1 and f_2, they are both join/split features (one of them can be a sequence feature), subgraph-flexible feature matching and their join (split) nodes are matched. If f_1 is a subgraph of f_2 (vise-verse), they are detected as additional branches, otherwise detected as different branches.*

5.2 Differences Related to Gateway-Flexible Feature Matching

This sub-section describes three types of differences that can be detected using gateway-flexible feature matching. Difference *different order* indicates that two features are with the same sets of activities but with different sets of edges or gateways. For example, the third pair of features in Figure 4 is detected as *different order*, because the orders of activities *a, b, c* are different in these features. Difference *different routing* indicates that two join/split features are with the same join/split activities and the same set of joined/split activities, but with different sets of edges or gateways. For example, the fourth pair of features in Figure 4 is detected as *different routing*, because the gateways are different in these two features. Difference *loop2once-off* indicates that one of two features is a loop feature and the other is not a loop feature, and these two features are with the same sets of activities, e.g., the last pair of features in Figure 4. Differences *different routing* and *loop2once-off* are special cases of difference *different order*. We do not consider differences repeatedly between features. That is if two features are detected as *different routing* or *loop2once-off* then they are not detected as *different order* anymore.

Definition 11 (Different Order). *Given two features f_1 and f_2, they are detected as different-order if they are gateway-flexible feature matching.*

Definition 12 (Different Routing). *Given two features f_1 and f_2, they are both join (split) features. They are detected as different-routing if and only if their split nodes are matched and they are gateway-flexible feature matching.*

Definition 13 (Loop2once-off). *Given two features f_1 and f_2, one is a loop feature and the other is not a loop feature. They are detected as loop2once-off if and only if they are gateway-flexible feature matching.*

6 Detecting Differences between Business Process Graphs

This section presents the steps to detect differences between business process graphs. There are totally three steps: activity matching, feature splitting and matching, and difference detecting.

Activities are compared firstly to find the mapping between activities of these two sets by analyzing their labels. Identical labels between activities of two models are mapped. For the rest of activities, each activity label in one model is compared with activity labels in the other model, using the metric defined in Definition 5. The comparison stops until it has a matching label or it finishes comparing with all labels in the other model without a matching. Take the pair of graphs in Figure 2 as an example. *Activity a, d, e and f* are identical activities between these two graphs; *activity b and b'* (*"Inbound Delivery Entered"* and *"Inbound Delivery Created"*) are interchanged activities; *activity c and g* are skipped activities.

The second step is to get all local maximal features. First, loop features are detected in a business process graph. Second, activities directly connected to gateways of a business process graph are marked. Then, the process graph is decomposed by duplicating the marked activities to get all local maximal sequence, split, and join features. Last, matching features of two process graphs are identified. Continuing with the example in Figure 2, no loop features in these graphs are found; *activity d* in *graph a* and *activity d, e* in *graph b* are marked. Then, *graph a* is decomposed into two join features, *F1.1, F1.2*; *graph b* is also decomposed into three join features, *F2.1, F2.2, F2.3*, as shown in Figure 2. Feature *F1.1* and feature *F2.1* are subgraph-flexible feature matching; Feature *F1.2* and feature *F2.2* are gateway-flexible feature matching.

Lastly, each pair of matched features is checked to detect differences defined in Section 5. For example, in Figure 2, Feature *F1.1* and feature *F2.1* are detected as *additional branches*; Feature *F1.2* and feature *F2.2* are detected as *different routing*.

Overall, for the process graphs in Figure 2, five differences are detected: two skipped activities, *activity c and g*; one pair of interchanged activities, *activity b and b'*; one *additional branches, feature F1.1 and F2.1*; one *different routing, feature F1.2 and F2.2.*

7 Evaluation

This section presents the evaluation of the technique proposed in this paper. Firstly, the experimental setup is described. Secondly, the experimental results are presented.

The experiments is made in the context of the SAP reference model. This is a collection of 604 business process models (described as EPCs) capturing business processes supported by the SAP enterprise system. On average the process models contain 21.6 nodes with a minimum of 3 and a maximum 130 nodes. The average size of node labels is 3.8 words. Similarities of pairs among 604 business process models are computed using the process similarity search technique in [24]. Pairs with similarity higher than 0.5 (more than 140,000 pairs) are selected for evaluating the difference detection technique proposed in this paper. All the experiment are run on a laptop with the Intel Core 2 Duo processor T7500 CPU (2.2GHz, 800MHz FSB, 4MB L2 cache), 4 GB DDR2 memory, the Windows 7 operating system and the SUN Java Virtual Machine version 1.6.

We run the experiment five times with all these selected process model pairs using our difference detection technique and the average execution time is 89ms for one pair, while in the worst case it takes 5.8s. The statistics of differences detected from these pairs are shown in Table 1. The cutoff value of label matching used in the experiment is 0.8. As expected, there are a lot of skipped activities while only a small amount of interchanged activities. This is because in the SAP reference models, the same functions (events) use the same label and different labels usually differ a lot. There are some cases of different and additional branches. Few other differences exist in these model pairs. This is because there are too many skipped activities and few features can have sets of the same (interchanged) activities. From Table 1 we can know that all the differences defined in this paper exists in the real-life process models. Therefore, the technique presented in this paper can provide valuable feedback about differences between process models.

Table 1. Statistics of differences between process models

Skipped Activities	4,823,217
Interchanged Activities	138
Additional Branches	447
Different Branches	1210
Different Order	12
Different Routing	82
Loop2once-off	54

8 Related Work

The work presented in this paper is related to three different types of techniques: business process difference detection, business process alignment, and business process retrieval.

Existing researches about difference detection follow into two branches: detecting syntactic or semantic differences between process models. Li et al. [17] and Küster et al. [15] focus on structures of business process models and proposes techniques to identify syntactic differences between these models. However, these techniques focus on how to edit one model to the other and the differences are detected based on nodes and edges. In this paper, we focus on the workflow patterns (features). This provides more syntactical meanings and can contain multiple differences based on nodes and edges in one difference based on features. On the other hand, Dijkman summarizes semantic differences [2] and proposes a technique based on trace equivalence to detect these differences [4].

Process alignment is complimentary to process difference detection. The former techniques align the matching parts and the latter techniques detect the mismatching parts between process models. Weidlich et al. propose a framework for process alignment in [21]. Weidlich et al. [19,20] also provide techniques to align models for the same process at different abstraction level based on the semantics of these models. Dijkman et al. [6] propose to align process models based on the syntax of models.

Another relevant research topic is process retrieval. Process difference detection and alignment techniques focus on comparing a pair of process models; while process retrieval techniques focus on comparing one given query model with a collection of process models. There are basically two branches of process retrieval: process similarity search [5,7,24,8] and process querying [1,12,11,25]. The method used in this paper is similar to the methods in [5,7,24,25]. There are two major differences besides different applications (this paper solves difference-detection instead of retrieval) are as follows. In [5,7] the whole process models are compared based on graph edit distance, while in this paper features (fragments) are used instead. In [24,25] similar features are used, however, feature matching is strict, i.e., both activities and flows must be matching. This cannot help identify differences, therefore, flexible matching is used in this paper instead.

9 Conclusion

This paper presents a technique for detecting syntactic differences between business process models using flexible feature matching. The technique provides valuable feedback about differences between process models with rich syntactic meanings based on workflow patterns (sequences, splits, joins, and loops). The experiment shows that all types of differences proposed in this paper exist in real-life process models.

There are some possible improvements on the technique described in this paper, which are left for future work. First, given two process models, the syntactic differences proposed in this paper can be used to efficiently edit one process model based on the other process model. Second, activity matching is mainly based on label similarity. However, similar activities can be labeled differently, e.g., synonyms, different levels of verbosity. Therefore, we applied more advanced metrics for label similarity that consider synonyms [4] and domain ontologies [13]. Third, the technique in this paper mainly focuses on activities and control-flows. However, process models often contain more information, e.g., resources and data used, which are also useful for detecting differences between models.

References

1. Awad, A.: BPMN-Q: A language to query business processes. In: Proceedings of EMISA 2007, Nanjing, China, pp. 115–128 (2007)
2. Dijkman, R.: A classification of differences between similar Business Processes. In: EDOC 2007, p. 37 (2007)
3. Dijkman, R.: Feedback on Differences between Business Processes. BETA Working Paper WP-234, Eindhoven University of Technology, The Netherlands (2007)
4. Dijkman, R.: Diagnosing differences between business process models. In: Dumas, M., Reichert, M., Shan, M.-C. (eds.) BPM 2008. LNCS, vol. 5240, pp. 261–277. Springer, Heidelberg (2008)
5. Dijkman, R., Dumas, M., García-Bañuelos, L.: Graph Matching Algorithms for Business Process Model Similarity Search. In: Dayal, U., Eder, J., Koehler, J., Reijers, H.A. (eds.) BPM 2009. LNCS, vol. 5701, pp. 48–63. Springer, Heidelberg (2009)
6. Dijkman, R., Dumas, M., Garcia-Banuelos, L., et al.: Aligning business process models. In: IEEE International Conference on Enterprise Distributed Object Computing, EDOC 2009, pp. 45–53. IEEE (2009)
7. Dijkman, R., Dumas, M., Van Dongen, B., et al.: Similarity of business process models: Metrics and evaluation. Information Systems 36(2), 498–516 (2011)
8. Dijkman, R.M., van Dongen, B.F., Dumas, M., et al.: A Short Survey on Process Model Similarity. In: Seminal Contributions to Information Systems Engineering - 25 Years of CAiSE, pp. 421–427 (2013)
9. van Dongen, B.F., Dijkman, R., Mendling, J.: Measuring Similarity between Business Process Models. In: Bellahsène, Z., Léonard, M. (eds.) CAiSE 2008. LNCS, vol. 5074, pp. 450–464. Springer, Heidelberg (2008)
10. Gao, X., Chen, Y., Ding, Z., et al.: Process Model Fragmentization, Clustering and Merging: An Empirical Study. In: BPM Workshops: The 4th International Workshop on Process Model Collection: Management and Reuse, Beijing, China (2013)
11. ter Hofstede, A.H.M., Ouyang, C., La Rosa, M., Song, L., Wang, J., Polyvyanyy, A.: APQL: A process-model query language. In: Song, M., Wynn, M.T., Liu, J. (eds.) AP-BPM 2013. LNBIP, vol. 159, pp. 23–38. Springer, Heidelberg (2013)
12. Jin, T., Wang, J., Wu, N., La Rosa, M., ter Hofstede, A.H.M.: Efficient and Accurate Retrieval of Business Process Models through Indexing. In: Meersman, R., Dillon, T.S., Herrero, P. (eds.) OTM 2010. LNCS, vol. 6426, pp. 402–409. Springer, Heidelberg (2010)
13. Ehrig, M., Koschmider, A., Oberweis, A.: Measuring similarity between semantic business process models. In: Proceedings of the 4th Asia-Pacific Conference on Conceptual Modelling, Ballarat, Victoria, Australia, pp. 71–80 (2007)
14. Kunze, M., Weidlich, M., Weske, M.: Behavioral Similarity - A Proper Metric. In: Rinderle-Ma, S., Toumani, F., Wolf, K. (eds.) BPM 2011. LNCS, vol. 6896, pp. 166–181. Springer, Heidelberg (2011)
15. Küster, J.M., Gerth, C., Förster, A., Engels, G.: Detecting and resolving process model differences in the absence of a change log. In: Dumas, M., Reichert, M., Shan, M.-C. (eds.) BPM 2008. LNCS, vol. 5240, pp. 244–260. Springer, Heidelberg (2008)
16. La Rosa, M., Reijers, H.A., van der Aalst, W.M.P., Dijkman, R.M., Mendling, J., Dumas, M., Garcia-Banuelos, L.: APROMORE: An advanced process model repository. Expert Systems with Applications 38(6), 7029–7040 (2011)

17. Li, C., Reichert, M.U., Wombacher, A.: On Measuring Process Model Similarity based on High-level Change Operations. In: Proceedings of the 27th International Conference on Conceptual Modeling, Barcelona, Spain, pp. 248–264 (2008)
18. Lu, R., Sadiq, S.K.: On the Discovery of Preferred Work Practice through Business Process Variants. In: Parent, C., Schewe, K.-D., Storey, V.C., Thalheim, B. (eds.) ER 2007. LNCS, vol. 4801, pp. 165–180. Springer, Heidelberg (2007)
19. Weidlich, M., Barros, A., Mendling, J., Weske, M.: Vertical Alignment of Process Models – How Can We Get There? In: Halpin, T., Krogstie, J., Nurcan, S., Proper, E., Schmidt, R., Soffer, P., Ukor, R. (eds.) BPMDS 2009 and EMMSAD 2009. LNBIP, vol. 29, pp. 71–84. Springer, Heidelberg (2009)
20. Weidlich, M., Weske, M., Mendling, J.: Change propagation in process models using behavioral profiles. In: SCC 2009, pp. 33–40. IEEE (2009)
21. Weidlich, M., Dijkman, R., Mendling, J.: The ICoP Framework: Identification of Correspondences between Process Models. In: Pernici, B. (ed.) CAiSE 2010. LNCS, vol. 6051, pp. 483–498. Springer, Heidelberg (2010)
22. Yan, Z., Grefen, P.W.P.J.: A Framework for Business Process Model Repositories. In: BPM Workshops: First International Workshop on Process in the Large (IW-PL 2010), Hoboken, NJ, USA (2010)
23. Yan, Z., Dijkman, R.M., Grefen, P.W.P.J.: Business Process Model Repositories - Framework and Survey. Information and Software Technology 54(4), 380–395 (2012)
24. Yan, Z., Dijkman, R.M., Grefen, P.W.P.J.: Fast Business Process Similarity Search. Distributed and Parallal Databases 30(2), 105–144 (2012)
25. Yan, Z., Dijkman, R., Grefen, P.: FNet: An Index for Advanced Business Process Querying. In: Barros, A., Gal, A., Kindler, E. (eds.) BPM 2012. LNCS, vol. 7481, pp. 246–261. Springer, Heidelberg (2012)

A Model-Checking Based Approach to Robustness Analysis of Procedures under Human-Made Faults

Naoyuki Nagatou[1,2] and Takuo Watanabe[1]

[1] Department of Computer Science, Graduate School of Information Science and Engineering, Tokyo Institute of Technology.
2-12-1 Oookayama, Meguro-ku, Tokyo, 152-8552, Japan
[2] PRESYSTEMS Inc., Kamimuzata 1461, Togane, Chiba, 283-0011, Japan
{nagatou,takuo}@acm.org

Abstract. We propose a model-checking approach for analyzing the robustness of procedures that suffer from human-made faults. Many procedures executed by humans incorporate fault detection and recovery tasks to recover from human-made faults. Examining whether such recovery tasks work as expected is crucial for preserving the trust and reliability inherent in safety-critical domains. To achieve this, we used a type of fault injection method that injects a set of human-made faults into a fault-free model of a given procedure; the fault set is selected according to Swain's discrete action classification. We use a model checker to determine paths to error states within the model and its properties formalized via CCS and LTL. We show the effectiveness of our method by investigating the recoverability of a real-world procedure.

Keywords: Linear temporal logic, process algebra, procedure, model checking, human-made fault, dependability, robustness.

1 Introduction

Humans use procedures in many scenarios such as industrial plants, aircrafts, and hospitals. Procedures in safety-critical domains typically require high levels of dependability, which is ensured not only by the skills and knowledge of domain experts but also by the inherent robustness of the given procedures, because domain experts may still make mistakes. In this paper, we present a formal approach for analyzing procedures for robustness in terms of human-made faults and logical characteristics of robust procedures.

A Hazard and Operability (HAZOP) study[1] is a structured and systematic examination of a planned process to identify and evaluate problems that may represent risks to the person or equipment. HAZOP was initially developed to analyze chemical plants, then later extended to study procedure, human and software. The examination in HAZOP is based on guide words such as NO OR NOT, MORE, LESS, AS WELL AS, PART OF, REVERSE and OTHER THAN. An HAZOP study starts by defining the scope and objectives of a system. The

C. Ouyang and J.-Y. Jung (Eds.): AP-BPM 2014, LNBIP 181, pp. 117–131, 2014.

system is divided into parts, then parameters of the parts are defined. Each guide word is applied to each of the parts and parameters for identifying deviation from design intent. Results are recorded on HAZOP worksheets, which are matrices with parts and parameters, and guide words.

If systems and their objectives are written in Petri net and a formal executable language in model checking, then results of each examination are executable. Execution is helpful to identify consequences, causes, and protection and detection mechanisms. Workflow net[2] is a subclass of a Petri net that has two special places: input place and output place. It is straightforward to map a procedure to a Workflow net. Automatical analysis is also available, for example to use LTSA[3].

Fields[4] applies model checking to the analysis of faults in human-computer interaction design. A model written in Murphi language[5] describes the interactions between a device and user's tasks performing in a usage. Combining a device model and a usage model as a user tasks represents a situation in which a user uses a device. User performance that deviates from design intent is injected into a combined model based on deviations that he defines. Deviations are coded as transition rules. The analysis of erroneous actions is performed in an injected model, and a property of hazard states or goal conditions. Another approach is Ezekiel and Lomuscio's approach[6]. They apply to distributed systems using a different fault model from Fields's fault model.

When a model checking tool reports counterexamples to hazard states, we need to understand what counterexamples represent. The logical characteristics of robust procedures understands the meaning of counterexamples, although the characteristic is not clear in the above. Numerous researchers have studied the characteristics of fault-tolerance systems[7,8,9]. Such research requires a fault model for modeling incorrect behavior against design intent and strategies for building fault-tolerant systems. Therefore, we introduce formalization of a fault model called discrete action classification (DAC)[10] and a strategy that includes error detection and error recovery, assuming that those tasks are embedded in a procedure.

As in many approaches using formal method, this study first requires a formal description of systems or procedures. Procedures are specified as a set of agent expressions in process algebra, i.e., Calculus of Communicating Systems (CCS)[11], and interactions between agents. An element in a procedure corresponds to an agent, and interactions between agents represent activities of humans and systems. The formal semantics of the DAC fault model are specified as translation rules on agent expressions. Recovery techniques do not explicitly appear in this framework because we assume that detection and recovery tasks are embedded within each procedure.

The verification framework uses properties to represent goals of a procedure. The properties intuitively state that given a goal property φ, for all possible path φ is satisfied. If a model with injected faults no longer satisfies a goal property, then we say that the procedure is not robust with regard to injected faults.In this paper, we present a validation check of a procedure in a hospital as an example.

1. When a patient enters to the blood sampling room, the patient gives a clerk in the room his/her ID-card.
2. The clerk has the receipt system read the ID card.
3. The system gives the clerk two receipt sheets and gives a nurse labels.
4. The clerk gives the patient one of the two receipt sheets; the clerk gives a nurse the other.
5. The nurse prepares tubes.
6. The nurse has the system read the labels.
7. The nurse puts the labels on the tubes.
8. The system calls a receipt ID.
9. The nurse confirms the name on the labels with a name on the receipt sheet.
10. The nurse receives a receipt sheet from the patient after the patient sits down.
11. The nurse compares the names on the receipt sheets.
12. The nurse extracts blood samples from the patient.
13. The nurse sends the blood samples to a lab.

Fig. 1. Example workflow for collecting blood samples in a hospital

We construct two distinct scenarios involving a single patient and two patients, then investigate the recoverability of the procedure.

Let us compare the study in this paper with process discovery and verification of event logs. Process discovery is a map from event logs to process models, and different from our study although related to. While in [12], observed event sequences are verified, our study verifies unobserved (not logged) sequences predicted by hazard analysis.

In addition to this introductory section, we organize this study as follows. Section 2 describes a motivating example for this study. Section 3 presents the formal frameworks (i.e., process algebra and linear temporal logic (LTL) , which are used in [13]). Section 4 provides a formalism of procedures in process algebra. Section 5 provides a fault model for human errors. Section 6 provides a notion of robustness and recovery, and describes how we investigate the robustness of procedures by model checking. In the last section, we summarize the contributions of our study and present avenues for future work.

2 Motivating Example

Fig. 1 shows an example workflow that describes the steps required for obtaining blood samples from a patient in a hospital. The workflow starts with a patient presenting his/her ID card, after he or she arrives in a room for blood sample extraction. A clerk makes a receipt system read the ID card. The system checks the patient's ID number in a data base, and if it corresponds to the ID number of the extraction request, the system prints two receipt sheets and some labels. The clerk receives the sheets and gives one of the sheets to the patient; the other sheet and the labels are given to a nurse. The nurse then prepares tubes and applies the labels to the tubes. The nurse also confirms the labels and makes the system read the labels. If an ID number on the labels is different than that on the sheet, then the nurse corrects the labels. The system displays the ID-number on an electrical board to call the patient. The patient then takes his/her seat and hands the sheet to the nurse. The nurse asks the patient his/her name and compares the patient's name with the name on a terminal. If these names are identical, then the nurse extracts blood into the tubes. The nurse takes both the blood samples and the sheets to a laboratory as the final step.

The designer of a procedure (typically, a supervisor) will adjust it when there are incidents, for example mistakes of patients. Steps 9 and 11 are introduced, depending on previous mistakes. These two redundant checks may prevent nurses from mistaking one patient for another, although there is no guarantee. A pseudo-execution of a procedure is helpful in validating the procedure. Therefore, in this paper, we discuss a framework of validating a procedure using model checking.

2.1 Human-Made Fault Classification

Procedures are performed by humans, although humans make mistakes. Human-made faults are essentially human actions exceeding limits of acceptability defined by a situation in which a human performs an action. Human performance can be affected by many factors, such as fatigue, external distractions, insufficient skill sets, a lack of motivation regarding correct or optimal performance, expectations and pressures that the performer senses from others, and a work situation that may not match an individual's abilities. These factors may cause humans to perform faulty actions or omit a crucial step of a procedure. While it is difficult to identify all faulty actions, such actions have certain identifiable features in this study, we identify these features.

We focus on unintentional faults committed by experts. On the basis of attention (performance) level[14, pp.53–96], such unconscious faults are classified into the following three categories: skill-based faults, rule-based faults and knowledge-based faults. This classification is useful for psychological research but not for robustness analysis because these refer to human variables. We are interested in human outputs as a result to be yielded, while the classifications in [14, pp.53–96] mention internal processes of the mind. A human perceives external inputs, processes such inputs, and produces outputs, which influence the success of the given tasks and procedures.

DAC[10] is a classification of human-made faults based on incorrect human outputs. The classification is organized as follows:

- Faults of Omission:
  - omits entire task
  - omits a step in a task
- Faults of Commission:
  - Selection fault: selects wrong control, mispositions a control, or issues the wrong command and/or information
  - Sequence fault: performs a task out of order

- Timing fault: performs a task too early or too late
- Qualitative fault: gives over performance or under performance than the proper or expected degree (too much or too little)

An omission fault is a fault in which one forgets to perform the entire task or a step of the task. In this study, we assumed that all tasks are atomic; because of this simplification, we ignored the "omits a step of a task" category. A commission fault is a fault in which one performs an incorrect action. Commission faults are classified into the four abovementioned categories. However, this paper focuses only on selection and sequence faults, because our modeling language[13] does not have a notion of time or quality.

3 Preliminaries

In this section, we present a subset of CCS[11] to express procedures, and a subset of Linear Temporal Logic (LTL) to express properties required by the procedures.

3.1 Process Algebra

CCS is one of process algebras that is based on communication between agents. Communication causes two types of actions: input and output actions. An input action is labeled by a name a and has a corresponding output action. An output action is identified by a bijective function $\overline{a}$. We call $\overline{a}$ of name a a co-name. We assume that a set of co-names is disjoint from the set of names.

If E and F are agents and a is a name,

- output prefix $(\overline{a}(e_1, \cdots, e_n) : E)$ is an agent,
- input prefix $(a(x_1, \cdots, x_n) : E)$ is an agent,
- summation of E and F, written E+F, is an agent,
- composition of E and F, written $E \mid F$, is an agent,
- constant $P(x_1, \cdots, x_n)$ is an agent given by $P(x_1, \cdots, x_n) \overset{\text{def}}{=} E$, and
- conditional `if` (b) (E) (F) is also an agent.

In the above, $x_1, \cdots$ are variables over a fixed set of values $\mathtt{V}$, $e_1, \cdots$ are expressions over $\mathtt{V}$, and b is a boolean expression over $\mathtt{V}$. We use $ZERO$ as a significant agent that does nothing.

Next, we describe the behavior of each agent via value passing. Consider $(a(x_1, \cdots, x_n) : E)$, where if $i \neq j$, then x_i is a different symbol from x_j. This contains offering values at a, and a transforms $(a(x_1, \cdots, x_n) : E)$ into E, whose behavior depends on $x_1, \cdots, x_n$, where $x_1, \cdots, x_n$ are free variables in E. We next introduce the scope of variables before we explain of free variables. We say that a binds $x_1, \cdots, x_n$ to $v_1, \cdots, v_n$, respectively, and their scopes are in E. If variables are not bound, they become free variables. We write the substitution of variables $x_1, \cdots, x_n$ to $[v_1/x_1, \cdots, v_n/x_n]$, for example, $(a(x_1, \cdots, x_n) : E)$

$$\frac{\langle E, k, env\rangle \overset{\alpha}{\rightarrow} \langle E', k, env\rangle}{\langle E{+}F, k, env\rangle \overset{\alpha}{\rightarrow} \langle E', k, env\rangle} \qquad \frac{\langle F, k, env\rangle \overset{\alpha}{\rightarrow} \langle F', k, env\rangle}{\langle E{+}F, k, env\rangle \overset{\alpha}{\rightarrow} \langle F', k, env\rangle}$$

$$\frac{\langle E, k, env\rangle \overset{\alpha}{\rightarrow} \langle E', k, env\rangle}{\langle E \mid F, k, env\rangle \overset{\alpha}{\rightarrow} \langle E', k(\lambda x.x \mid F), env\rangle} \qquad \frac{\langle F, k, env\rangle \overset{\alpha}{\rightarrow} \langle F', k, env\rangle}{\langle E \mid F, k, env\rangle \overset{\alpha}{\rightarrow} \langle F', k(\lambda x.E \mid x), env\rangle}$$

$$\frac{\langle E, k, env[v_1/x_1, \cdots, v_n/x_n]\rangle \overset{\alpha}{\rightarrow} \langle E', k, env[v_1/x_1, \cdots, v_n/x_n]\rangle}{\langle P(v_1, \cdots, v_n), k, env\rangle \overset{\alpha}{\rightarrow} \langle E', k, env[v_1/x_1, \cdots, v_n/x_n]\rangle} \quad (P(x_1, \cdots, x_n) \overset{\text{def}}{=} E)$$

Fig. 2. Operational semantics of the process calculus

becomes $E[v_1/x_1, \cdots, v_n/x_n]$. We can then write the inference rule for input prefix as

$$\frac{}{\langle a(x_1, \cdots, x_n) : E, k, env\rangle \overset{a(v_1, \cdots, v_n)}{\rightarrow} \langle E, k, env[v_1/x_1, \cdots, v_n/x_n]\rangle},$$

where k and env are a continuation and an environment, respectively.

We next consider $\overline{a}(e_1, \cdots, e_n) : E$. $\overline{a}$ demands values of expressions. $\overline{a}$ always enables an experiment. Let v_i be a value to evaluate e_i for $1 \leq i \leq n$, we can then give an inference rule for output prefix as

$$\frac{}{\langle \overline{a}(e_1, \cdots, e_n) : E, k, env\rangle \overset{\overline{a}(v_1, \cdots, v_n)}{\rightarrow} \langle E, k, env[(v_1, \cdots, v_n)/a]\rangle},$$

where $eval$ is an evaluator; for all $1 \leq i \leq n$, v_i is $eval(e_i, env)$.

Next, we consider a conditional agent with two arms. Let α be an element of the union set of names and co-names. Evaluating b to one of the boolean values {`true`,`false`}, we can give an inference rule for conditional agents as

$$\frac{E \overset{\alpha}{\rightarrow} E'}{\texttt{if (true)}\ (E)\ (F) \overset{\alpha}{\rightarrow} E'} \qquad \frac{F \overset{\alpha}{\rightarrow} F'}{\texttt{if (false)}\ (E)\ (F) \overset{\alpha}{\rightarrow} F'}.$$

The other behaviors of agents are defined as shown in Fig. 2.

3.2 Linear Temporal Logic

We use LTL to describe goal properties of procedures. We first assume that a trace has initial states and is a finite sequence of states. We write the length of trace $\sigma = s_0\, s_1 \cdots s_n$ to $|\sigma|$ in which $|\sigma|$ is $n + 1$. We write the suffix of $\sigma = s_0\, s_1 \cdots s_i \cdots s_n$ starting at i as $\sigma^{i..} = s_i \cdots s_n$, and the ith state as σ^i.

We assume a vocabulary $x, y, z, \cdots$ of variables for data values. For each state, variables are assigned to a single value. A state formula is any well-formed first-order formula constructed over the given variables. Such state formulas are evaluated on a single state to a boolean value. If the evaluation of state formula p becomes true over s, then we write $s[p] = \texttt{tt}$ and say that s satisfies p, where `tt` and `ff` are truth values, denoting *true* and *false* respectively. Let φ and ψ be temporal formulas, a temporal formula is inductively constructed as follows:

- a state formula is a temporal formula,
- the negation of a temporal formula $\neg\varphi$ is a temporal formula,
- $\varphi \vee \psi$ and $\varphi \wedge \psi$ are temporal formulas, and
- $\Box\,\varphi$, $\Diamond\,\varphi$, $\circ\,\varphi$, and $\varphi\,\mathcal{U}\,\psi$ are temporal formulas.

We next define the semantics of temporal formulas over a trace. If trace σ satisfies property φ, then we write $\sigma \models \varphi$. Furthermore,

- if p is a state formula, then $\sigma \models p$ iff $\sigma^0[p] = \mathtt{tt}$ and $|\sigma| \neq 0$.
- $\sigma \models \neg\varphi$ iff $\sigma \not\models \varphi$,
- $\sigma \models \varphi \vee \psi$ iff $\sigma \models \varphi$ or $\sigma \models \psi$,
- $\sigma \models \varphi \wedge \psi$ iff $\sigma \models \varphi$ and $\sigma \models \psi$,
- $\sigma \models \Box\varphi$ iff for all $0 \leq i < |\sigma|$, $\sigma^{i..} \models \varphi$,
- $\sigma \models \Diamond\varphi$ iff there exists $0 \leq i < |\sigma|$ such that $\sigma^{i..} \models \varphi$,
- $\sigma \models \circ\,\varphi$ iff $\sigma' \models \varphi$ where $\sigma' = \sigma$ if $|\sigma| = 1$ and $\sigma' = \sigma^{1..}$ if $|\sigma| > 1$,
- $\sigma \models \varphi\,\mathcal{U}\,\psi$ iff there exists $0 \leq k < |\sigma|$ s.t. $\sigma \models \psi$ and for all $j < k$, $\sigma \models \varphi$.

A formula φ is satisfiable if there exists a sequence σ such that $\sigma \models \varphi$. Given set of traces T and formula φ, φ is valid over T if for all $\sigma \in T$, $\sigma \models \varphi$.

3.3 Relationships between Agents and Formulas

In this subsection, we describe the relationship between algebraic models and LTL formulas. The modeling language enables us to pass values via input prefix $\alpha(e)$ and output prefix $\overline{\alpha}(x)$ with the same name. Execution of $\alpha(e)$ produces value v of e. Execution of $\overline{\alpha}(x)$ causes a single assignment to x. Furthermore, the execution of two actions causes atomic assignment $x := v$, that is, communication between two agents produces a new state by changing the values of the variables. This is similar to the first paragraph in Section 3.3 of [15, page 290].

This atomic assignment changes states, and we represent the change as $s[v/x]$, which denotes a change in the values of x in s to v. A state is a mapping from variables to values. Assuming that $\mathtt{Var_E}$ is a set of variables that appears in prefixes in agent E with range $\mathtt{V}$, $s : Var_E \rightarrow \mathtt{V}$. For example, the evaluation $s[x = y]$ of $x = y$ at s becomes $s[x] = s[y]$, and at $s[v/x]$, $s[v/x][x] = s[v/x][y]$, i.e., $v = s[y]$.

Therefore, communication between agents produces a sequence of assignments, which then produces a sequence of state changes called a trace. Let a set of traces produced by agent E be T. If for all traces $\sigma \in T$, $\sigma \models \varphi$, then we state that φ is valid over E and write $E \models \varphi$.

4 Workflow Model

Before providing a model of human-made faults, we characterize a procedure as a set of sequences of tasks, where a task has a principal and an object. A performance of a task is the transmission of information, in which the information is transmitted in the form of messages. Messages cause send-actions in principals and receive-actions in objects. In addition, tasks have dependencies on one another. A dependency is indicated by an ordering of tasks. Tasks, principals, objects and message are just symbols that are assigned to actual meanings depending on skill, knowledge, cognitive processes, etc.

Formally, we characterize a procedure using three tuple $(\mathcal{C}, \mathcal{T}, \mathcal{O})$, which we will call a workflow model. $\mathcal{C}$ is a set of principals and objects, $\mathcal{T}$ is a set of tasks or messages, and $\mathcal{O}(\subseteq \mathcal{T} \times \mathcal{T})$ is a set of orders of two tasks. For example, given the procedure for extracting blood samples in a hospital (from Fig. 1 above), we extract the following sets that make up the workflow model:

- $\mathcal{C}$ ={patient, nurse, label, receipt, clerk, tube, laboratory},
- $\mathcal{T}$ ={"give a clerk ID card", "make system read ID card", "give clerk two sheets", "give receipt to nurse",$\cdots$},
- $\mathcal{O}$ ={("give a clerk ID card", "make system read ID card"), ("make system read ID card", "give clerk two sheets"),$\cdots$}.

More specifically, $\mathcal{C}$ and $\mathcal{T}$ are sets of symbols, each symbol is assigned to an element of domain $\mathcal{D}$ by interpretation $\mathcal{I}: \mathcal{C} \cup \mathcal{T} \rightarrow \mathcal{D}$. $\mathcal{D}$ and $\mathcal{I}$ denote parts of cognitive processes, skills and knowledge of humans.

4.1 Mapping Procedures via Process Algebra

We map procedures to agents using the process algebra described in Section 3.1 above. The given mapping method is based on algorithm 1 of [16]. The mapping produces agents corresponding to principals and objects of tasks; each element of $\mathcal{C}$ becomes an agent. For example, $Patient, Nurse, Receipt, Clerk, Tube$ and Lab are all agents.

Moreover, the mapping produces a co-name and a name for each task because the performance of a task causes a send-action and a receive-action. The mapping produces co-names in agents for principals and names in agents for objects. For example, as shown in part in Fig. 3 for "give receipt to nurse" $\in \mathcal{T}$, output prefix $\overline{give\_receipt\_to\_nurse}(e)$ appears in $Clerk$ and input prefix $give\_receipt\_to\_nurse(e)$ appears in $Nurse$, where e is an expression over a set of values.

The order of occurrence of names and co-names preserves the order between pairs of tasks. If, for example, ("give receipt to nurse","give label"),("give label","prepare tube") $\in \mathcal{O}$, then we produce

$$give\_receipt\_to\_nurse(receipt\_id\_Nurse, name\_on\_receipt\_Nurse) : \\ \overline{give\_label}(receipt\_id\_on\_label\_Nurse, name\_on\_label\_Nurse) : \\ \overline{prepare\_tube} : \cdots .$$

An overall procedure is represented by the composition of all agents. For example,

$$W \stackrel{\text{def}}{=} (Patient \mid Nurse \mid Clerk \mid Receipt \mid Tube \mid Lab).$$

$$
\begin{aligned}
Nurse \stackrel{\text{def}}{=} \; & (give\_receipt\_to\_nurse(receipt\_id\_Nurse, name\_on\_receipt\_Nurse) : \\
& give\_label(receipt\_id\_on\_label\_Nurse, name\_on\_label\_Nurse) : \\
& \overline{prepare\_tube} : \\
& \overline{make\_system\_read\_label} : \\
& \overline{put\_label\_on\_tube(name\_on\_label\_Nurse)} : \\
& confirm\_label(patient\_name\_Nurse) : \\
& receive\_receipt(name\_on\_receipt\_for\_patient\_Nurse) : \\
& compare\_receipt(patient\_name\_Nurse) : \\
& \overline{extract\_blood} : \\
& \overline{take\_sample\_to\_lab}(name\_on\_receipt\_for\_patient\_Nurse, \\
& \qquad\qquad name\_on\_label\_Nurse) : \\
& Nurse) \\
Clerk \stackrel{\text{def}}{=} \; & (give\_id\_card(patient\_name\_on\_card\_Clerk) : \\
& \overline{make\_system\_read\_card}(patient\_name\_on\_card\_Clerk) : \\
& give\_receipt(receipt\_id\_Clerk, name\_on\_receipt\_Clerk) : \\
& \overline{give\_receipt\_to\_patient}(receipt\_id\_Clerk, name\_on\_receipt\_Clerk) : \\
& \overline{give\_receipt\_to\_nurse}(receipt\_id\_Clerk, name\_on\_receipt\_Clerk) : \\
& Clerk)
\end{aligned}
$$

Fig. 3. Sample workflow model for error-free tasks regarding the blood extraction procedure of Fig. 1

A formal definition of agent $Nurse$ is shown in Fig. 3. Note that the definition allows for recursion. Furthermore, we may be able to define actions in opposite direction. For example, $compare\_name\_on\_receipt$ and $confirm\_name$ have the opposing direction (i.e., from objects to principals). So, we may choose the object-to-principal direction to simplify the model although we always use the principal-to-object direction.

Given workflow model $\mathcal{W}$ as $(\mathcal{C}, \mathcal{T}, \mathcal{O})$, we map each element of $\mathcal{W}$ to each element of the process algebra as follows. $\mathcal{C}$ becomes a set of constant agents for principals and objects. Given a set of names N and a set of co-names $\overline{N}$, $\mathcal{T}$ becomes $N \cup \overline{N}$. $\mathcal{O}$ becomes the order of action prefixes.

In this study, we do not mention skills, knowledge, and cognitive processes of humans because these are hidden by domain $\mathcal{D}$ and interpretation $\mathcal{I}$. We assume that all task sequences before injecting faults reach the goal state of a procedure.

5 Injection of Human-Made Faults

We formalize human-made faults, except timing faults, because the process algebra and logic that we use do not have a notion of time. The formalism produces agents for tasks in which human-made faults may occur in opposition to the correct meaning of tasks mentioned in Section 4. In this study, we treat omission faults, selection faults, and sequence faults as human-made faults.

Given set of faulty tasks H and workflow model W, let a set of names in W be N and a set of co-names be $\overline{N}$. We next describe the injection of omission faults. Let $t \in H$ and $H \subseteq \mathcal{T}$. For each $t_1, t_2 \in \mathcal{T}$, if $(t_1, t), (t, t_2) \in \mathcal{O}$, then $\{(t_1, t_2)\}$ is added to $\mathcal{O}$. We therefore obtain $\mathcal{W}_H = (\mathcal{C}, \mathcal{T}, \mathcal{O}')$, where $\mathcal{O}' = \mathcal{O} \cup \{(t_1, t_2)\}$. For faulty tasks in a procedure, we construct constant agents for them. Summation agents are constructed as follows: let an action for t be $\alpha \in N \cup \overline{N}$,

$$(\alpha(e) : ZERO) + (ZERO).$$

We next describe the injection of selection faults. Given two sets of tasks $H \subseteq \mathcal{T}$ and $H' \not\subseteq \mathcal{T}$. Let $t \in H$, $t' \in H'$. If $(t_1, t), (t, t_2) \in \mathcal{O}$, then we replace t by t'. We then obtain (t_1, t') and (t', t_2). Adding (t_1, t') and (t', t_2) to $\mathcal{O}$, $\mathcal{O}' = \{(t_1, t'), (t', t_2)\} \cup \mathcal{O}$, and $\mathcal{T}' = H' \cup \mathcal{T}$. We then obtain $W_{H,H'} = (\mathcal{C}, \mathcal{T}', \mathcal{O}')$. Let actions for faulty tasks be $\alpha(e_1)$ and $\alpha'(e')$. Using the summation operator, we construct the following agent:

$$\alpha(e_1) : ZERO + \alpha'(e') : ZERO.$$

Faulty tasks with omission and selection faults can be represented by the following single agent:

$$(\alpha(e_1) : ZERO) + (\alpha'(e') : ZERO) + (ZERO).$$

Finally, we describe the injection of sequence faults. Let $t_1, t_2 \in H$, then $\mathcal{O}$ becomes $\mathcal{O}' = \{(t_1, t_2), (t_2, t_1)\} \cup \mathcal{O}$. We can then obtain $\mathcal{W}_H = (\mathcal{C}, \mathcal{T}, \mathcal{O}')$. For tasks with faults, we construct constant agents for them in a workflow model. Let agents for the faulty tasks be $Task\_t1$ and $Task\_t2$, and actions be $\alpha_1(e_1), \alpha_2(e_2) \in N \cup \overline{N}$, we then have

$$Task\_t1 \stackrel{\text{def}}{=} (\alpha_1(e_1) : ZERO),$$
$$Task\_t2 \stackrel{\text{def}}{=} (\alpha_2(e_2) : ZERO).$$

Using composition, we construct the following agent:

$$Task\_t1 \mid Task\_t2.$$

For tasks with omission, selection and sequence faults, we construct the following agent:

$$Task\_t1 \stackrel{\text{def}}{=} (\alpha_1(e_1) : ZERO + \alpha_1'(e_1') : ZERO + ZERO),$$
$$Task\_t2 \stackrel{\text{def}}{=} (\alpha_2(e_2) : ZERO + \alpha_2'(e_2') : ZERO + ZERO),$$

and $Task\_t1 \mid Task\_t2$.

Using these formalisms, we inject human-made faults into the model shown in Fig. 3. We suppose $H = \{$"confirm label", "compare receipt"$\}$. Our formalism produces the following two agents:

$Task\_confirm\_label$ and
$Task\_compare\_receipt$.

If nurses make mistakes in these tasks, then composition of those agents replaces $confirm\_label$ and $compare\_receipt$ in $Nurse$ shown in Fig. 3. Fig. 4 shows these injected agents. We obtain multiple traces from the agents and a single trusted service trace from the agents in Fig. 3.

$$
\begin{aligned}
&\mathit{Task\_confirm\_label} \stackrel{\mathrm{def}}{=} \\
&\quad (\mathit{confirm\_label}(\mathit{name\_on\_display\_TaskConfLabel}) : \mathit{ZERO} \\
&\quad +\mathit{ZERO}) \\
&\mathit{Task\_compare\_receipt} \stackrel{\mathrm{def}}{=} \\
&\quad (\mathit{receive\_receipt}(\mathit{name\_on\_receipt\_for\_patient\_TaskCompReceipt}) : \\
&\qquad \mathit{compare\_receipt}(\mathit{name\_on\_display\_TaskCompReceipt}) : \mathit{ZERO} \\
&\quad +\mathit{ZERO}) \\
&\mathit{Nurse1} \stackrel{\mathrm{def}}{=} \\
&\quad (\overline{\mathit{extract\_blood}} : \\
&\quad \overline{\mathit{take\_sample\_to\_lab}}(\mathit{name\_on\_receipt\_for\_patient\_Nurse1}, \\
&\qquad \mathit{name\_on\_label\_Nurse1}) : \mathit{Nurse}) \\
\mathit{Nurse} \stackrel{\mathrm{def}}{=} &\ (\mathit{give\_receipt\_to\_nurse}(\mathit{receipt\_id\_Nurse}, \mathit{name\_on\_receipt\_Nurse}) : \\
&\quad \mathit{give\_label}(\mathit{receipt\_id\_on\_label\_Nurse}, \mathit{name\_on\_label\_Nurse}) : \\
&\quad \overline{\mathit{prepare\_tube}} : \\
&\quad \overline{\mathit{make\_system\_read\_label}} : \\
&\quad \overline{\mathit{put\_label\_on\_tube}}(\mathit{name\_on\_label\_Nurse}) : \\
&\quad (\mathit{Task\_confirm\_label} \mid \mathit{Task\_compare\_receipt} \mid \mathit{Nurse1}))
\end{aligned}
$$

Fig. 4. Fault-injected model of nurse's tasks

6 Robustness Analysis

In this section, We use model checking to investigate the existence of traces that cause an irrecoverable state based on human-made faults. In Section 6.1 below, we provide characteristics of robust procedures. Next, in Section 6.2, we apply this framework to the example model given in Fig. 4.

6.1 Characterization of Robust Procedures

Robustness is an attribute that systems are often required to have. This attribute depends on external factors, which are human-made faults in this study. These human-made faults cause error states. Therefore the robustness of a procedure affected by human-made faults is defined as follows.

Definition 1 (Robustness). *Let e be an error state formula, $\hat{e}$ an error state, and φ a formula. If error state $\hat{e}$ exists on σ and $(\sigma, \hat{e}) \models \Diamond\varphi$, then σ is robust on error e, and we say that σ is (e, φ)-robust. If φ is valid over all traces obtained from W, then W is (e, φ)-robust.*

This definition intuitively means that a sequence of tasks σ reaches a goal, which satisfies φ from an error state.

We next focus on a technique for guaranteeing robustness, which consists of detection of and recovery from errors. Error detection identifies an error state and Error recovery changes an error state to a state without any detected errors. A recovery task typically occurs after detection of an error state, and the task is performed in various steps. We therefore define recovery as follows.

Definition 2 (Recovery). *Given σ, if $\sigma \models \Box(\neg e) \vee \Diamond(e \wedge \Diamond r)$, then we say that σ is weakly recoverable, wherein e and r are an error state formula and a recovery state formula, respectively.*

The definition intuitively means that a recovered state appears after an error state appears.

We abbreviate $\Box(\neg e) \vee \Diamond(e \wedge \Diamond r)$ as $e \rhd r$. The following lemma intuitively means that recovered state formula r_i becomes true after error state formula e_i is true.

Lemma 1. *Given σ, let e_i be an error state formula and r_i a recovery state formula, if $\sigma^{\hat{r_i}\cdots} \models (\bigwedge_i r_i) \wedge \Diamond\varphi$ and $\sigma \models \bigwedge(e_i \rhd r_i)$ then $\sigma \models (\bigwedge_i(e_i \rhd r_i)) \ni \Diamond\varphi$*

Proof Sketch *Let e_i be an error state formula, r_i a recovery state formula. We suppose that $\sigma \models (e_i \rhd r_i)$ holds over σ.*

$$\begin{aligned} &\models (e_i \rhd r_i) \\ \textit{iff} &\models \Box(\neg e_i) \vee \Diamond(e_i \wedge \Diamond r_i) \\ \textit{iff} &\models \Box(\neg e_i) \ \textit{or} \ \models \Diamond(e_i \wedge \Diamond r_i) \end{aligned}$$

In the case that $\Diamond(e_i \wedge \Diamond r_i)$ holds, there exists s_i such that e_i and $\Diamond r_i$ hold. Let all r_i hold at s_j where $j \geq i$, and we suppose that $(\sigma, s_j) \models (\bigwedge_j r_j) \ni \Diamond\varphi$. Then, there exists s_k such that φ holds. Therefore, $\sigma \models (e_i \rhd r_i) \ni \Diamond\varphi$. Supposing for all e_i, $\sigma \models (e_i \rhd r_i) \ni \Diamond\varphi$, $\bigwedge_i \sigma \models (e_i \rhd r_i) \ni \Diamond\varphi$.

In the other case in which $\neg e_i$ always holds, $\Diamond\varphi$ clearly holds because we suppose that a procedure before injections of errors indeed reaches its goal. □

By Lemma 1, if there exists σ such that $\sigma \models \neg\Diamond\varphi$, then there exists e_i such that $\sigma \models \neg(e_i \rhd r_i)$. More specifically, there are errors that cannot be recovered. Thus, workflow model W that includes σ is not (e_i, φ)-robust. Therefore, giving the negation, we will verify recoverability of procedures by model checking.

6.2 Model Checking

A designer of the procedure shown in Fig. 1 above may believe that the procedure prevents nurses from mistaking patients because of two checkpoints, namely step 9 to confirm and step 11 to compare. Those checkpoints are fundamental to detect error. Therefore, we set $H = \{$"confirm label", "compare name"$\}$, with Fig. 4 illustrating the injected model. We assume that the rest of the tasks do not cause human-made faults.

One of goals of this procedure is for nurses to take tubes and sheets without making any mistakes. Nurses guarantee that the blood of patients is not mixed up, or more specifically, they guarantee that the names on labels match those on the sheets. When those names have the same value, the procedure is complete. Given an agent for a laboratory

$$\begin{aligned} Lab \stackrel{\text{def}}{=} \ & take\_sample\_to\_lab(name\_on\_receipt\_Lab, name\_on\_label\_Lab) : \\ & Lab, \end{aligned}$$

$$
\begin{aligned}
&\mathit{Receipt}(\mathit{receipt\_id\_Receipt}) \stackrel{\text{def}}{=} \\
&\quad \mathit{make\_system\_read\_card}(\mathit{patient\_name\_Receipt}) : \\
&\quad \overline{\mathit{give\_receipt}}(\mathit{receipt\_id\_Receipt}, \mathit{patient\_name\_Receipt}) : \\
&\quad \overline{\mathit{give\_label}}(\mathit{receipt\_id\_Receipt}, \mathit{patient\_name\_Receipt}) : \\
&\quad \mathit{make\_system\_read\_label} : \\
&\quad \overline{\mathit{call\_receipt\_id}}(\mathit{receipt\_id\_Receipt}) : \\
&\quad \overline{\mathit{confirm\_label}}(\mathit{patient\_name\_Receipt}) : \\
&\quad \overline{\mathit{compare\_receipt}}(\mathit{patient\_name\_Receipt}) : \\
&\quad \mathit{Receipt}(\mathit{receipt\_id\_Receipt} + 1), \\
&\mathit{Clerk}() \stackrel{\text{def}}{=} \\
&\quad \mathit{give\_id\_card}(\mathit{patient\_name\_on\_card\_Clerk}) : \\
&\quad \overline{\mathit{make\_system\_read\_card}}(\mathit{patient\_name\_on\_card\_Clerk}) : \\
&\quad \mathit{give\_receipt}(\mathit{receipt\_id\_Clerk}, \mathit{name\_on\_receipt\_Clerk}) : \\
&\quad \overline{\mathit{give\_receipt\_to\_patient}}(\mathit{receipt\_id\_Clerk}, \mathit{name\_on\_receipt\_Clerk}) : \\
&\quad \overline{\mathit{give\_receipt\_to\_nurse}}(\mathit{receipt\_id\_Clerk}, \mathit{name\_on\_receipt\_Clerk}) : \mathit{Clerk}, \\
&\mathit{Patient}(\mathit{name\_Patient}) \stackrel{\text{def}}{=} \\
&\quad \overline{\mathit{give\_id\_card}}(\mathit{name\_Patient}) : \\
&\quad \mathit{give\_receipt\_to\_patient}(\mathit{receipt\_id\_Patient}, \mathit{name\_on\_receipt\_Patient}) : \\
&\quad \mathit{call\_receipt\_id}(\mathit{called\_id\_Patient}) : \\
&\quad \overline{\mathit{ask\_name}}(\mathit{name\_on\_receipt\_Patient}) : \\
&\quad \mathit{extract\_blood} : \mathit{ZERO}, \\
&\mathit{Tube}(\mathit{id\_Tube}) \stackrel{\text{def}}{=} \\
&\quad \mathit{prepare\_tube} : \\
&\quad \mathit{put\_label\_on\_tube}(\mathit{name\_on\_label\_Tube}) : \mathit{Tube}(\mathit{id\_Tube} + 1).
\end{aligned}
$$

Fig. 5. Definition of error-free agents

We now identify property

$$\Box\Diamond(\mathit{name\_on\_receipt\_Lab} = \mathit{name\_on\_label\_Lab}). \quad (1)$$

Next, we investigate the recoverability of procedures using a model checking tool[13]. We give the negation of (1), as in

$$\Diamond\Box\neg(\mathit{name\_on\_receipt\_Lab} = \mathit{name\_on\_label\_Lab}),$$

and instances of the injected model to the tool. An instance represents a situation in which a procedure works. When one nurse and two patients appear, the instance becomes

$$\mathit{Patient}(\text{"}\mathit{Alice}\text{"}) \mid \mathit{Patient}(\text{"}\mathit{Bob}\text{"}) \mid \mathit{Nurse} \mid \mathit{Clerk} \mid \mathit{Receipt} \mid \mathit{Tube}(\mathit{id}) \mid \mathit{Lab}.$$

Definitions of $\mathit{Receipt}, \mathit{Clerk}, \mathit{Patient}$ and Tube are shown in Fig. 5. In this case, a failure is possible, i.e., tubes may be mistaken. Three instances of patient names exist in this procedure: (a) the name that the nurse has, (b) the name that a patient has, and (c) the name on the label. As described in Fig. 1, checkpoint 9 compares a name on the label with the name that the nurse has, and checkpoint

Table 1. Result of model checking

patients	workspace	Failure
one	—	not appear
two	not isolate	appear
two	isolate	not appear

11 compares names that the nurse and the patient have. If one of these checkpoints fail, then the procedure cannot guarantee that those names are the same. Thus, the nurse following this procedure may mistake blood samples of patients.

The next case we consider is the situation in which we divide workspaces per patient. This situation is represented with two tube instances, and the instance of the workflow model is as follows:

$$Patient(\text{"}Alice\text{"}) \mid Patient(\text{"}Bob\text{"}) \mid Nurse \mid Clerk \mid Receipt \mid Tube(\text{"}For\_Alice\text{"}) \mid Tube(\text{"}For\_Bob\text{"}) \mid Lab.$$

The definition of $Tube$ is as follows:

$$Tube(id\_Tube) \stackrel{\text{def}}{=} prepare\_tube : put\_label\_on\_tube(name\_on\_label\_Tube) : ZERO.$$

In this case, a failure does not appear, even if the nurse slips two check activities.

The last case is a simple situation in which there is one patient. The instance of the workflow model becomes as follows:

$$Patient(\text{"}Alice\text{"}) \mid Nurse \mid Clerk \mid Receipt \mid Tube(id) \mid Lab.$$

In this case, a failure does not appear even if the two check points fail.

We therefore obtain the results shown in Table 1. We conclude that "confirm label" and "compare name" are not sufficient to avoid making a mistake in the procedure, and the procedure is not E, φ-robust, where E ={omits of "confirm label" and "compare name"} and $\varphi = \Diamond\Box\neg(name\_on\_receipt\_Lab = name\_on\_label\_Lab)$.

7 Conclusion

We have discussed characteristics of robust procedures and proposed a verification technique using model checking for procedures that may suffer from human-made faults. The discussion provided a framework for the recoverability of procedures that have been injected with human-made faults. As an example, we investigated the recoverability of a procedure for taking blood samples.

We formalized human-made faults using process algebra and separated them into the following five categories: omission, selection, sequence, timing and qualitative faults. Our formalism treats omission, selection, and sequence faults. However, we did not attempt to handle timing faults and qualitative faults. Thus, for future work, we intend to incorporate a notion of time and formalize qualitative faults.

References

1. IEC 61882:2001: Hazard and operability studies (HAZOP studies)– Application guide. IEC, Geneva (2001)
2. van der Aalst, W.M.P.: Verification of workflow nets. In: Azéma, P., Balbo, G. (eds.) ICATPN 1997. LNCS, vol. 1248, pp. 407–426. Springer, Heidelberg (1997)
3. Karamanolis, C.T., Giannakopoulou, D., Magee, J., Wheater, S.M.: Model checking of workflow schemas. In: Proceedings of the 4th International Conference on Enterprise Distributed Object Computing, EDOC 2000, pp. 170–181. IEEE Computer Society, Washington, DC (2000)
4. Fields, R.E.: Analysis of Erroneous Actions in the Design of Critical Systems. PhD thesis, University of York (January 2001)
5. Formal Verification Group, School of Computing, University of UTAH: Murphi Model Checker, `http://www.cs.utah.edu/formal_verification/Murphi`
6. Ezekiel, J., Lomuscio, A.: A methodology for automatic diagnosability analysis. In: Dong, J.S., Zhu, H. (eds.) ICFEM 2010. LNCS, vol. 6447, pp. 549–564. Springer, Heidelberg (2010)
7. Krishnan, P.: A semantic characterisation for faults in replicated systems. Theoretical Computer Science 128(1-2), 159–177 (1994)
8. Bernardeschi, C., Fantechi, A., Gnesi, S.: Model checking fault tolerant systems. Software Testing, Verification and Reliability 12(4), 251–275 (2002)
9. Gnesi, S., Lenzini, G., Martinelli, F.: Logical specification and analysis of fault tolerant systems through partial model checking. In: Etalle, S., Mukhopadhyay, S., Roychoudhury, A. (eds.) Proceedings of the International Workshop on Software Verification and Validation (SVV 2003), Mumbai, India. Electronic Notes in Theoretical Computer Science, vol. 118, pp. 57–70. Elsevier, Amsterdam (2003)
10. Swain, A.D., Guttmann, H.E.: Handbook of Human Reliability Analysis with Emphasis on Nuclear Power Plant Applications. Draft Report NUREG/CR-1278, U.S. Nuclear Regulatory Commission Office of Nuclear Regulatory Research, Washington, DC (May 1982)
11. Milner, R.: Communication and Concurrency. Prentice-Hall, Inc., Upper Saddle River (1989)
12. van der Aalst, W.M.P., de Beer, H.T., van Dongen, B.F.: Process mining and verification of properties: An approach based on temporal logic. In: Meersman, R., Tari, Z. (eds.) OTM 2005. LNCS, vol. 3760, pp. 130–147. Springer, Heidelberg (2005)
13. PRESYSTEMS Inc.: A model checker: NHK, `http://www4.ocn.ne.jp/~presys/index_en.html`
14. Reason, J.: Human Error. Cambridge University Press (1990)
15. Lamport, L., Schneider, F.B.: The "Hoare Logic" of CSP, and all that. ACM Transactions on Programming Languages and Systems (TOPLAS) 6(2), 281–296 (1984)
16. Puhlmann, F.: Soundness verification of business processes specified in the pi-calculus. In: Meersman, R., Tari, Z. (eds.) OTM 2007, Part I. LNCS, vol. 4803, pp. 6–23. Springer, Heidelberg (2007)

Incorporating the Effects of Organizational Structure into Business Process Simulation

Jinyoun Lee[1], Sung-Nyun Hearn[2], Jinwoo Kim[1], and Injun Choi[1,*]

[1] Dept. of Industrial and Management Engineering
Pohang University of Science and Technology (POSTECH)
San 31, Hyoja-dong, Pohang, Republic of Korea
{ljyboy,phantazy,injun}@postech.ac.kr
[2] Samsung Electronics Co., Ltd.
Suwon, Gyeonggi, Republic of Korea
sn.hearn@samsung.com

Abstract. Today's organizations change or redesign business processes and/or organizational structures, more frequently than ever before, to innovate and adapt to the rapidly changing environment. Business process simulation attempts to provide more effective prediction of the effects of process and organizational redesign. Most existing approaches consider only business processes, however, without considering organizational structures that can significantly affect business process performance. This paper introduces an approach to incorporating the effects of organizational structure into business process simulation. First, the paper identifies various factors for designing organizational structures that can affect business process performance. Then, it proposes a simulation model that combines business processes and organizational structures to show how the above factors can be incorporated into a simulation model. The proposed approach will enable more precise prediction of the changes caused by process and organizational redesign.

Keywords: Business process simulation, Business process reengineering, Business process analysis, Organizational structure, Redesign.

1 Introduction

Today's organizations change or redesign business processes and/or organizational structures, more frequently than ever before, to innovate and adapt to the rapidly changing environment. Business process reengineering (BPR) and process innovation (PI), which are the most recognized efforts for process redesign, have not produced desired results [1-4]. Hence, the concept of business process simulation is proposed to provide more effective prediction of the effects of process redesign [5-8]. Most existing approaches consider only business processes, however, without considering organizational structures that can significantly affect business process performance [5, 9, 10].

* Corresponding author.

C. Ouyang and J.-Y. Jung (Eds.): AP-BPM 2014, LNBIP 181, pp. 132–142, 2014.

This paper introduces an approach to incorporating the effects of organizational structure into business process simulation. First, the paper identifies various factors for designing organizational structures that can affect business process performance. Then, it proposes a simulation model that combines horizontal business processes and vertical organizational structures to show how the above factors can be incorporated into a simulation model. The proposed approach will enable more precise prediction of the changes caused by process and organizational redesign.

The remainder of the paper is organized as follows. Section 2 discusses related research. Section 3 describes the effects of organizational structure on business process performance. Section 4 proposes an approach for deriving a business process simulation model which incorporates the organizational effects. Section 5 concludes the paper.

2 Related Work

Many organizations frequently make use of simulation techniques for business process studies to innovate and adapt to the rapidly changing environment. In many BPR and PI projects, business analysts simulate redesigned processes to validate the processes and identify possible problems caused by the changes in processes [5, 8, 11, 12].

Since business processes are executed on organizational structures in practice, the structures can significantly affect the business process performance [9, 10, 13, 14]. In particular, even when the same business processes are executed, performance can differ depending on the organizational structures. Thus, only integrated simulation approach that includes both the business processes and the organizational structures can realistically facilitate analysis, design and redesign of organizations [5].

Most existing approaches, however, did not consider the organizational structures when they simulated business processes [5, 10, 14, 15, 16]. Moreover, they only concentrated on how to optimize business processes while relatively little attention has been placed on how organizational structures should be changed according to the results of business process simulation [5, 10].

There have been some attempts to overcome the limitation of previous business process simulation research efforts. Russell, N et al. (2004) discussed the impact of organizational resources to consider them in the modeling of business processes [17]. Chen, C. and Tsai, C. (2008) addressed the theoretical gaps between BPR and organizational restructuring (OR) in organizational change, and proposed the 'process reengineering-oriented organizational change exploratory simulation system' (PROCESS) for facilitating organizational change of BPR and OR simultaneously [11]. However, they have considered only a partial perspective of organizational structures such as organizational resources and departments. Furthermore, they have not addressed how incorporating the effects of organizational structure into the business process simulation.

To complement the previous research, this paper identifies various factors for designing organizational structures that can affect business process performance. Then, it proposes an improved business process simulation model that incorporates the effects of organizational structure into a simulation model.

3 The Effects of Organizational Structure on Business Process Performance

This section discusses essential organization design theories, especially, departmentalization and centralization, which are the two major concepts that affect the business process performance. This section explains how they can be represented in a business process simulation.

An organizational structure is defined as "the sum total of the ways in which an organization divides its labor into distinct tasks and then achieves coordination among them" [18]. In terms of organizational structure, tasks which require related skills, authorities, and responsibilities are grouped into jobs [19]. Furthermore, jobs are grouped together to coordinate common tasks, leading to the assembly of departments. This process is known as departmentalization [20]. Departments can be created according to certain criteria including function, product, geography, and etc., and these criteria also determine the type of the organizational structure.

After the departments are created, an administrative hierarchy of authority is established for coordination between departments [18]. A certain level of the administrative hierarchy is assigned to have the authority to make decisions, and this process is referred to as centralization [15]. In general, high-level managers such as the CEO approve and direct all tasks in highly centralized organizations and lower-level managers or employees are delegated the authority and responsibility in organizations centralized relatively in low degree.

This paper focuses on the effects of departmentalization and centralization on business process performance.

3.1 The Effect of Departmentalization

The effect of departmentalization can be shown as the number of transfers of work between departments for the execution of business processes. In terms of business processes, a transfer of work is performed when a task is finished to the next task in the process. In terms of organizational structure, however, the transfer of work is carried from the department where the task is executed to the department where the next task will be executed. The departmentalization criterion affects transfers of work. A department transfers work to other departments to execute business processes. The number of transfers of work between departments may be different depending upon the organizational structures that execute the same business process. Hence, the transfers of work between departments may take a long time in an organization while the same transfers can take significantly less time in other organizations. This effect of departmentalization should be addressed in the business process simulation model.

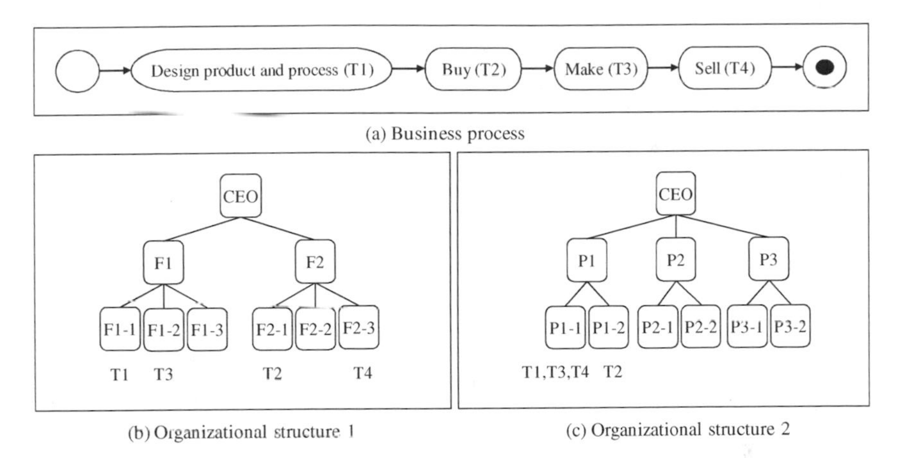

(a) Business process

(b) Organizational structure 1

(c) Organizational structure 2

Fig. 1. Example of a business process and two different organizational structures

Figure 1 illustrates a business process and two organizational structures. The process is a part of "Create physical asset {Manufacturer}" as defined in the MIT Process Handbook [21]. In the figure, the process is represented by the activity diagram in Unified Modeling Language (UML). The activity diagram has frequently been used to model business processes [22]. The organizational structures consist of departments grouped by function (Figure 1(b)) and product (Figure 1(c)). The departments responsible for each task are marked in the organizational structures. For example, the "Design product and process (T1)" task is performed in department "F1-1" within organizational structure 1 (Figure 1(b)) and in "P1-1" within organizational structure 2 (Figure 1(c)).

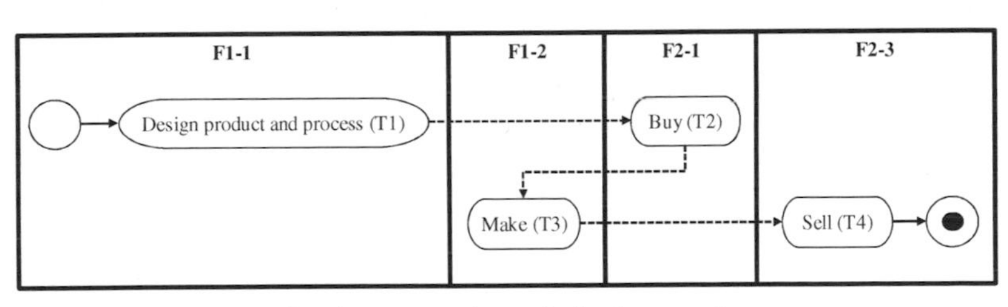

(a) Business process in organizational structure 1

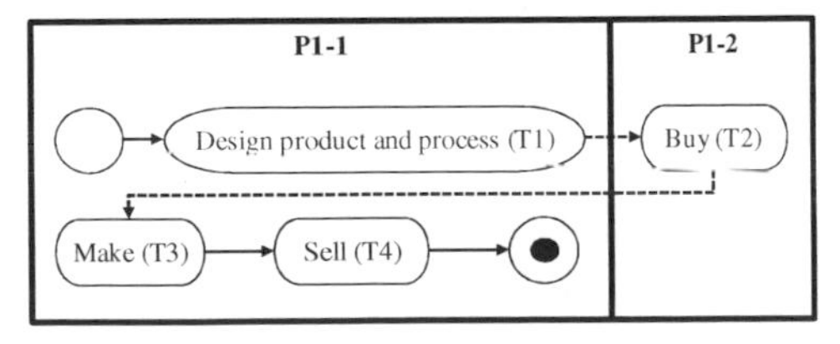

(b) Business process in organizational structure 2

Fig. 2. Business process executed in the organizational structures in Figure 1

Figure 2 illustrates the business process executed in each of Figure 1's organizational structures using activity diagrams with swim lanes. Both units responsible for the tasks and transfers of work between departments can be explicitly represented in the figure. Three transfers of work occur between departments in organizational structure 1 for performing the business process while two transfers of work occur between departments in organizational structure 2 for executing the same process, which is denoted by dotted arrows in the figure.

The transfer of work between departments requires more time than the transfer of work within a department due to factors such as:

- Handover of material, product, and related information between departments
- Different priorities, working cycles, and resource availabilities between departments
- Coordination between departments

Specially, conflicts can occur when coordinating departments due to a lack of communication and common goals (i.e., silo effect). The silo effect disrupts the flow and significantly decreases the business process performance.

3.2 The Effect of Centralization

The effect of centralization can be shown as the involvement of the administrative hierarchy in transfers of work between process tasks. Superiors have official authority and responsibility to coordinate subordinates. That is, the transfer of work between subordinates is managed by superiors with administrative tasks such as approval and direction. The degree of centralization determines the hierarchical level involved in the transfer of work. The involvement of administrative hierarchy can affect business process performance due to the time required to perform administrative tasks between superiors and subordinates. Further, delay time can occur because of the resource availability of a superior who manages many subordinates.

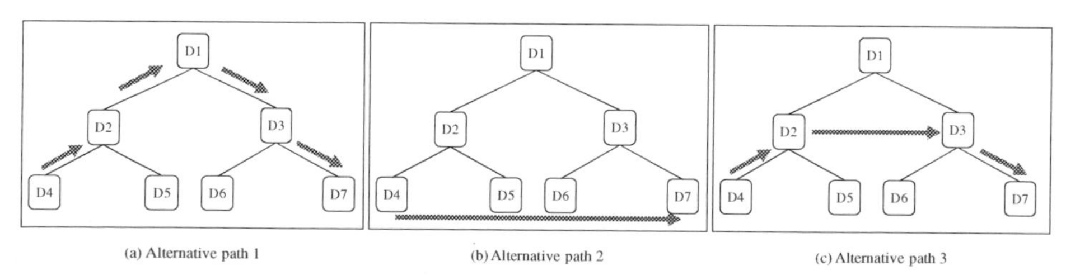

Fig. 3. Examples of paths for a transfer of work with different degrees of centralization

Figure 3 illustrates examples of administrative hierarchy involvement in the transfer of work between D4 and D7 with different degrees of centralization. In alternative path 1, the transfer of work is performed via authority relationships in the organizational structure with the highest degree of centralization. This is similar to "direct supervision" as presented by Mintzberg [18]. In alternative path 2, the transfer of work is directly executed without involving any supervisors with the lowest degree of centralization, which is referred to as "mutual adjustments" [18]. Alternative path 3 represents another possible path for the transfer of work using an intermediate degree of centralization.

For more precise simulations, the role of superiors involved in transfers of work should be considered. Superiors can directly perform transfers of work between them. Otherwise, their role can be limited to simply checking the status of processes that are being executed. Their roles determine the specific paths for the transfer of work within the organizational structure.

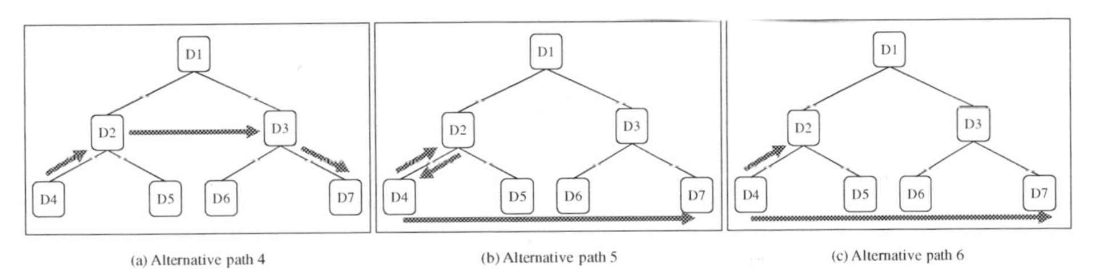

Fig. 4. Examples of transfer of work with different roles of superiors in the same degree of centralization

Figure 4 illustrates examples of paths in organizational structures for the transfer of work between D4 and D7 with different roles of superiors in the same degree of centralization. In alternative path 4, D2 and D3 directly perform the transfer of work. In alternative path 5, D4 performs the transfer of work after getting approval and direction from D2. In alternative path 6, D4 carries out the transfer of work while reporting to D2 (i.e., without getting approval and direction from D2).

The participants in a transfer of work should also be considered. Transfers of work can be performed solely by one department. For example, in alternative path 4, D2 can transfer work to D3. Transfers of work can also be jointly performed by two departments. For example, D2 and D3 can jointly coordinate the transfer of work. This should be clearly defined because the simulation results can be affected by resource availability.

4 Modeling of Business Process Simulations Integrating the Organizational Structure

A business process simulation that incorporates the organizational structure can be modeled through the following three steps: (1) modeling the business process, (2) modeling the organizational structure, and (3) modeling the business process simulation that incorporates the organizational structure.

First, the business process is modeled with its primary tasks and the execution order of the tasks to make products or provide services. Administrative tasks, such as coordination and approval, should not be included in this step because transfers of work between organizational units and the involvement of administrative hierarchy can differ depending on organizational structures. The example of the business process in Figure 5(a) consists of four tasks with three transfers of work between tasks.

When modeling the organizational structure, above all, the type of organizational structure should be defined. As described in section 2, departments can be created according to certain criteria that determine the type of organizational structure (e.g., functional structure, product structure, etc.). The administrative hierarchy is then built based on the departments. After designing the type of organizational structure, the degree of centralization should be determined. Furthermore, the specific role of superiors and participants in transfers of work should be considered. The example of the organizational structure in Figure 5(b) consists of nine departments and two levels of administrative hierarchy in which three transfers of work are performed between departments to execute the business process (from D4 to D6, D6 to D7, and D7 to D9). Among them, D2 and D3 are involved in the transfer of work between D6 and D7. D2 transfers work jointly with D3 after approving T2, and then D3 directs D7 to perform T3. Other transfers of work between departments are directly performed without involving any superior units.

The final step derives the business process simulation model by combining the business process and organizational structure defined above. The simulation model is defined as an activity diagram. The activity diagram can represent all the necessary information for the execution of the business process using factors such as stereotypes, attributes, notes, and constraints [22]. Thus, the simulation model can represent detailed information for a simulation such as the average time and distribution required to perform tasks.

A new simulation model incorporating the effects of organizational structure is introduced based on the business process model. First, the simulation model imports the primary tasks from the business process model with the responsible department. Subsequently, all types of transfers of work activities are inserted between the tasks to represent the transfers of work between departments. In this paper, the transfers of work activities are classified into three types to represent the pattern of the activities. The three types are defined as follows.

- Coordination activity (C): Horizontal coordination or handover between departments
- Approval activity (A): Hierarchical approval between superior and subordinate departments
- Direction activity (D): Hierarchical direction between superior and subordinate departments

The simulation model in Figure 5(c) is derived from the business process in Figure 5(a) and the organizational structure in Figure 5(b). It consists of five transfers of work activities: three coordination activities, one approval activity, and one direction activity. These are represented by dotted, rounded rectangles in the figure with transfer of work activities. The responsible departments are represented under the rectangle label. For example, D6 performs the approval activity to D2, and D2 and D3 jointly perform the coordination activity between them.

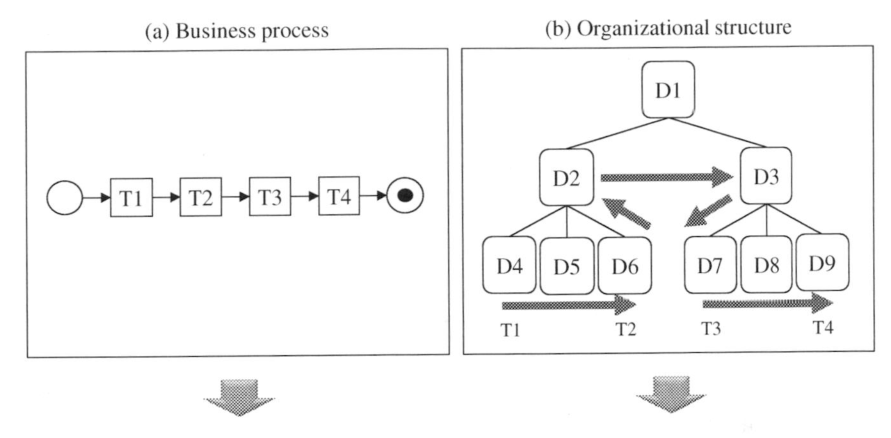

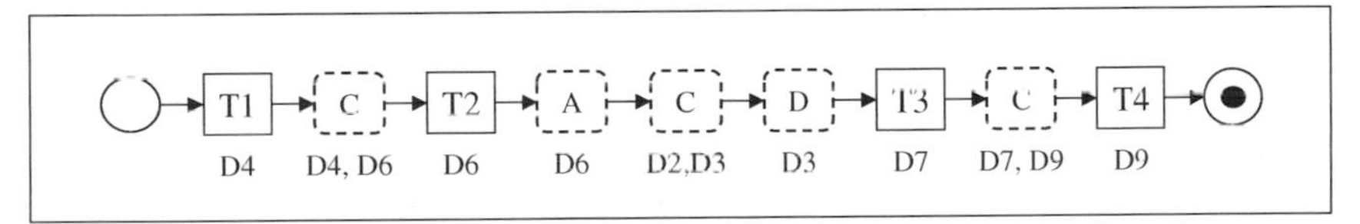

Fig. 5. Examples of a business process, an organizational structure, and a business process simulation model integrating the effects of organizational structure

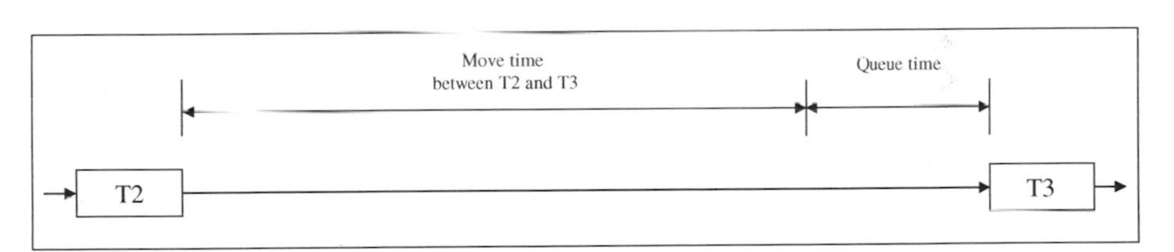

(a) Transfer of work time in existing business process simulation model

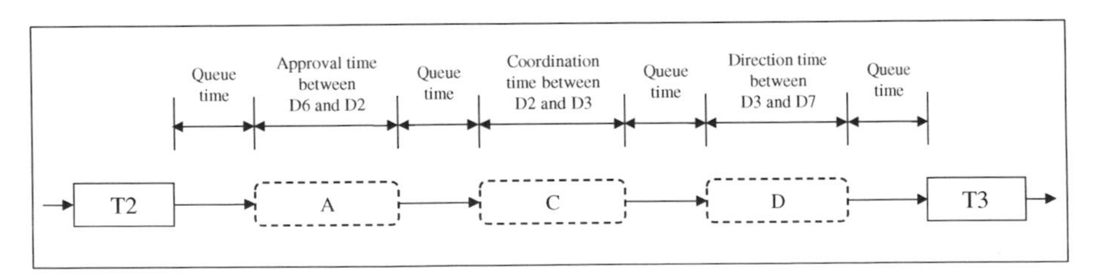

(b) Transfer of work time in proposed model

Fig. 6. The different perspective of the transfer of work time between the existing business process simulation model and the proposed one

The perspective of the transfer of work time and the manner of analysis are the biggest differences between the existing business process simulation model and the proposed one, and an example is illustrated in Figure 6. In the existing model, the transfer of work time is divided into only two types, which are 'move time' (the time it takes to move a case between tasks) and 'queue time' (the time a case spends waiting in a queue), not reflecting the effect of organizational structure (see Figure 6 (a)). In this case, the simulation results only reflected task time, and the transfer of work time was analyzed only in the perspective of tasks or business processes without considering the effects of organizational structure. On the other hand in the proposed model, the transfer of work time can be divided on three types of transfer of work time (approval time, coordination time, and direction time) and 'queue time' (see Figure 6(b)). In this case, the relationships between departments can be analyzed more precisely. For example, suppose that the transfer of work time between T2 and T3 is a critical problem resulting in poor business process performance and coordination time between D2 and D3 covers most of the transfer of work time between T2 and T3. In this case, the existing model will regard the relationship between D6 and D7 as the main reason of poor business process performance, whereas the proposed model will point out that the lack of communication between D2 and D3 can be a main reason. Therefore, the proposed model enables more precise analysis of the current situations and prediction of the changes caused by process and organizational redesign.

5 Conclusion

Even when the same business processes are executed, performance can differ depending on the organizational structures. This paper proposes a business process simulation model that incorporates the effects of organizational structure. First, it describes the effects of departmentalization and centralization on the execution of business processes, specifically highlighting the transfer of work between departments and the involvement of administrative hierarchy. Second, it proposes an approach for deriving a simulation model combining a horizontal business process and a vertical organizational structure. The simulation model is defined as an extended business process model to explicitly represent the effects of organizational structure.

Business processes can be analyzed and optimized more practically and meaningfully by the proposed simulation model since they are performed within organizational structures in practice. The proposed simulation model can reduce cost and risk, and also increase the success rates of BPR by pre-evaluating business processes and organizational structures. Furthermore, the proposed simulation model can support the design of appropriate organizational structures for business processes.

As an extension of this study, case studies should be performed to illustrate the importance and feasibility of the proposed approach. In addition, the proposed simulation model will adopt scheduling procedure between resources for transfer of work activities. For simpler and faster business process analysis and simulation, the tasks and transfer of work activities will be generalized into numerical values to make default parameters and thus to calculate the simulation results without specific data such as task time and transfer of work time between departments. In addition, the qualitative aspects like users' experience and skills (important aspects to reduce coordination and authority control) will be taken into account in the proposed model. Finally, a simulation tool will be developed for the automatic construction of simulation model with business processes and organizational structures.

Acknowledgements. This research was supported by the Industrial Strategic Technology Development Program (No. 10045047) through the National IT Industry Promotion Agency (NIPA) funded by the Ministry of Trade, Industry & Energy (MOTIE, Korea).

References

1. Ulmer, J., Belaud, J., Le Lann, J.: A Pivotal-Based Approach for Enterprise Business Process and ISIntegration. Enterprise Information Systems 7(1), 61–78 (2013)
2. Smith, H., Fingar, P.: Business process management: The third wave. Meghan-Kiffer Press, New York (2003)
3. Garvin, D.: The Processes of Organization and Management. Sloan Management Review 39(4), 33–50 (1998)
4. Goodstein, L.D., Butz, H.E.: Customer Value: The Linchpin of Organizational Change. Organizational Dynamics 27(1), 21–34 (1998)
5. Barjis, J., Verbraeck, A.: The relevance of modeling and simulation in enterprise and organizational study. In: Barjis, J. (ed.) EOMAS 2010. LNBIP, vol. 63, pp. 15–26. Springer, Heidelberg (2010)
6. van der Aalst, W.M.P.: Business process simulation revisited. In: Barjis, J. (ed.) EOMAS 2010. LNBIP, vol. 63, pp. 1–14. Springer, Heidelberg (2010)
7. Gregoriades, A., Sutcliffe, A.: A Socio-Technical Approach to Business Process Simulation. Decision Support System 45(4), 1017–1030 (2008)
8. Greasley, A.: Using Business-Process Simulation within a Business-Process Reengineering Approach. Business Process Management Journal 9(4), 408–420 (2003)
9. Hearn, S., Choi, I.: Creating a Process and Organization Fit Index: An Approach Toward Optimal Process and Organization Design. Knowledge and Process Management 20(1), 21–29 (2013)
10. Chen, C., Tsai, C.: Developing a Process Re-Engineering-Oriented Organizational Change Exploratory Simulation System (PROCESS). International Journal of Production Research 14(16), 4463–4482 (2008)
11. Giaglis, G.M., Paul, R.J., Hlupic, V.: Integrating Simulation in Organizational Design Studies. International Journal of Information Management 19(3), 219–236 (1999)
12. Clemons, E.K.: Using Scenario Analysis to Manage the Strategic Risks of Reengineering. Long Range Planning 28(6), 123–123 (1995)
13. Becker, J., Kugeler, M., Rosemann, M.: Process management: A guide for the design of business processes. Springer (2003)
14. Hammer, M., Stanton, S.: How Process Enterprises really Work. Harvard Business Review 77, 108–120 (1999)
15. Daft, R.: Organization Theory and Design (Thomson South-Western), 8th edn., Mason, OH (2004)

16. Russell, N., van der Aalst, W.M.P., ter Hofstede, A.H.M., Edmond, D.: Workflow resource patterns: Identification, representation and tool support. In: Pastor, Ó., Falcão e Cunha, J. (eds.) CAiSE 2005. LNCS, vol. 3520, pp. 216–232. Springer, Heidelberg (2005)
17. Mintzberg, H.: The Structuring of Organisations: A Synthesis of the Research. New Jersey (1979)
18. Lee, H.M.: Job analysis & redesign organization. Joongang Economy, Seoul (2009)
19. Robbins, S.P.: Organizational behavior, 11th edn. Pearson, New Jersey (2005)
20. MIT Process Handbook, `http://process.mit.edu/`
21. Jiang, P., Mair, Q., Newman, J.: The Application of UML to the Design of Processes Supporting Product Configuration Management. International Journal of Computer Integrated Manufacturing 19(14), 393–407 (2006)

Author Index